AN
ACTION
GUIDE TO
MAKING
QUALITY
HAPPEN

Robert Damelio
William Englehaupt

QUALITY RESOURCES ®

A Division of The Kraus Organization Limited

902 Broadway, New York, New York 10010

Most Quality Resources books are available at quantity discounts when purchased in bulk.
For more information contact:

Special Sales Department
Quality Resources
A Division of The Kraus Organization Limited
902 Broadway
New York, NY 10010
800-247-8519

Printed in the United States of America

99 98 97 96 95   10 9 8 7 6 5 4 3 2 1

Quality Resources
A Division of The Kraus Organization Limited
902 Broadway
New York, NY 10010
800-247-8519

The paper used in this publication meets the minimum requirements of American National Standard for Information Sciences—Permanence of Paper for Printed Library Materials, ANSI Z39.48–1984.

ISBN 0–527–76291–1

Library of Congress Cataloging-in-Publication Data

Englehaupt, Bill
    An action guide to making quality happen / Bill Englehaupt,
  Robert Damelio.
      p.   cm.
    ISBN 0–527–76291–1
    1. Total quality mamagement—Handbooks, manuals, etc. 2. Quality
  assurance—Management—Handbooks, manuals, etc. I. Damellio,
  Robert. II. Title.
  HD62.15.E54   1995
  658.5'62—dc20
                                                              95-1340
                                                              CIP

DEDICATION

To my mother, Ruth McCain, whose unwavering faith in me is a constant source of inspiration.

—Robert Damelio

To my Carrie for her love and support and my father, Bill—I know he would have been proud.

—William Englehaupt

TABLE OF CONTENTS

Acknowledgments

I wish to thank Ralph Damello, Wes Greenhill, Doug Morton, Ray Svenson, Geary Rummler, and Ken Wilburn, for the combined influence they have had on my life, and David, Rick, Pete, Albert, Richard, Chip, Hamo, Larry, Greg, Laurie, DonAnn, Niki, and Robin, for helping me understand what family and friendship is all about.

To the wonderful folks at the Irving Black-Eyed Pea who demonstrate their commitment to customer satisfaction with every meal. Thank-you for keeping my stomach well-fueled and my thirst quenched.

Finally, I'd like to thank my clients for allowing me the opportunity to be of service. Without you, none of this would be possible.

—Robert Damelio

I would like to personally acknowledge Robert, Robert, and Bill for starting me on the quality journey; Mike for sharing with me his love of processes; Phil for cheerleading; and Bev for helping me get started.

—William Englehaupt

Both authors wish to extend a special acknowledgment to Brian Joiner and John Domenick for allowing their material to be included in this book.

‖Introduction

SOME INITIAL ASSUMPTIONS

You'll find this book most helpful if:

- Your company is about to implement or is currently implementing a Quality Management System (QMS).

- You wish to do your part to help ensure that the implementation is successful.

- You are already aware of key quality improvement concepts.

QUALITY AND FITNESS IMPROVEMENT

We all know you don't become fit overnight. You make gradual progress one day at a time. However, if you follow a set of key principles, your level of fitness will measurably increase. The same is true with taking the actions needed to implement a QMS.

Throughout this book, we will periodically use the analogy of improving your level of fitness to illustrate how you can successfully implement a QMS. Why? First of all, the two present at least four similar challenges:

1. They both require behavior changes.

2. They are both based on a proven set of principles.

3. They both use data to set objectives, monitor progress, measure results, and get feedback.

4. They both take time to get lasting results.

BEHAVIOR CHANGES

The first similarity between fitness and quality improvement is that they both require changes in behavior. Consider the following statement: "You can change without improvement, but you can't improve without change."

Making quality happen means changing how some things get done. But how do you know what changes to make, and whether they will be improvements (changes for the better) or simply low-yield actions that, for whatever reason, seemed appropriate at the time? This book explains how by taking the right actions at the right time, you can make sure that the changes that accompany managing in a quality-conscious environment will result in improvements that you will experience personally as well as professionally.

PROVEN PRINCIPLES

The second similarity between fitness and quality improvement is that to achieve either, you must follow a set of proven principles. For example, it takes a combination of diet and exercise to increase your level of fitness.

Here are some of the proven principles that help you select actions that are appropriate to help you improve your level of fitness or improve the quality of a work process that you manage.

Fitness Principles	QMS Principles
• Eat foods made up of 20 percent fat or less.	• The voice of the customer should drive the business.
• If you burn off more calories than you take in, you will lose weight.	• All work is a process, and all processes contain variation.
• Regular exercise burns off calories.	• Reducing process variation reduces defects, cycle time, and costs.
• Some exercises burn off more calories than others in the same amount of time.	• Reducing defects and cycle time increases customer satisfaction.
• Selecting the type of exercise, the exercise schedule, duration, and the like are personal choices that should reflect your own preferences, constraints, fitness objectives, and lifestyle.	• Increased customer satisfaction and reduced costs increases profits.

BOTH REQUIRE DATA

How do you know if you are becoming more fit or whether quality is, in fact, improving? The answer is to collect data that is relevant to the process you wish to improve and to the performance targets associated with that process.

Here are some examples of the data that you might use:

Fitness data	Quality data
• Percent body fat	• Cycle time
• Calories; calories per gram of fat	• Defects per unit
• Pulse rate	• Customer satisfaction level
• Cholesterol level	• Return on quality

IT TAKES TIME TO SEE RESULTS

How long does it take for a person to become fit? One month? Six? The answer depends on a number of factors, such as the person's level of fitness to start with, their desired level of fitness, the rate at which they exercise, the amount of calories they consume, the nutritional mix of calories they consume, and so on—and that's for one person! Suppose you had a group of 10. How long would it take for all 10 to become fit?

Now, as a manager, ask yourself, "How long does it take to improve quality?" Actually, this question comes up in virtually every workshop on quality improvement conducted by the authors.

The short answer is, "About the same amount of time it would take if everyone in your group, department, or company set out to improve his or her level of fitness."

The challenge with both is in knowing the actions you can take daily in order to make regular progress. Neither quality nor fitness will result from crash programs.

Besides these four similarities, perhaps the most important reason we use the fitness analogy is that it is personal, and though it may have been difficult, chances are, you or someone you know has become fit. And, believe it or not, some companies have also dramatically improved quality. After all, nothing worthwhile is easy!

WHAT THIS BOOK IS AND ISN'T

This book *is not:*

- A rehash of the seven tools for continuous improvement.
- A step-by-step discussion of a continuous improvement process.
- About how to win the Baldrige Award.
- A philosophical discussion or explanation of any of the work of the quality gurus (i.e., Deming, Juran, Crosby, etc.)
- Long on theory.

Rather, this book is:

- A handbook for managers who want to make quality happen in their part of the business.
- Action-oriented.
- About proven tools and principles that you can use to improve quality daily.
- Designed to help you today!

WHY THIS BOOK? WHAT'S IN IT FOR YOU?

This book is for you if you have these typical concerns:

1. You are uncertain how quality improvement affects your role.
2. You believe that quality is the right thing to do, but you're not sure how to do it.
3. You wish to learn managerial practices that will increase customer satisfaction and improve quality for yourself, as well as how to better encourage and support the application of quality management practices by those with whom you work.
4. You are unsure how committed upper management is, or how long it may take them to get to your area, so you want to "get on with it" yourself.

Let's look at each of these typical concerns more closely.

THE MANAGER'S ROLE IN A QUALITY SYSTEM

For most people, the idea of a QMS is abstract and intangible, especially in the early stages of implementation. In Section 2, we describe a QMS as a set of related processes whose purpose is to increase customer satisfaction. We also view a quality management system as "a set of values that describe how to run the business every day." For now, there are two key points you should keep in mind. First, seen through the eyes of your associates, *you* are the QMS; that is, the perceptions that your associates form of the company's commitment to quality improvement are based to a very large extent on what *you* do and say as a manager. Thus, if your company has undertaken a quality improvement initiative, you are now in the spotlight whether or not you want to be there.

The second key point is that a QMS is not something unique to or intended solely for managers. Rather, the purpose of a QMS is to focus the collective efforts of everyone within the organization so that the level of satisfaction that internal and external customers experience will increase.

Section 2 of this book explains and details the seven elements or related processes that make up a QMS. You'll see that the manager's role is to establish or strengthen each of those elements and to ensure that the ties between each element and the work processes you manage are firmly in place.

QUALITY IS THE RIGHT THING TO DO, BUT I'M NOT SURE HOW TO DO IT

A second concern that many managers voice is that they are unsure of the specific actions they should take to improve quality. This is especially true of those managers who work in administrative, sales, finance, marketing, purchasing, or support functions, or who supervise scientific, or knowledge-workers.

Our work with managers has surfaced three recurring themes: First, everyone believes that quality is the right thing to do; second, almost everyone believes that quality is someone else's responsibility (usually the next level of management); and third, most managers are uncertain regarding the specific actions *they* should take to "make quality happen."

You'll find the focus of this book is on the specific actions *you* can take as a manager to successfully implement a QMS in your part of the business. For example, at the end of each chapter in Section 2, you will find a list of practical everyday actions you can take to implement each of the seven elements that make up a QMS. And in Section 3, we provide you with tools to assess your planned actions and analyze your current to-do lists in light of quality management system values so that you can be sure you walk the talk!

YOU WISH TO LEARN AND REINFORCE
QUALITY MANAGEMENT PRACTICES

If you're like most of the managers we work with, you are probably very busy to say the least, which leads directly to a third concern that managers often share with us. Sure, you want to improve quality and be a good corporate citizen, but can't we just give you the Cliff Notes version so you can focus in on just the important stuff? Okay. Section 1 is for you, especially Chapter 2 where we describe the three key managerial principles that support quality.

YOU WANT TO "GET ON WITH IT" YOURSELF

Finally, the fourth concern that high-achievers often state goes something like this, "If I have to wait until _____ gets here, I'll be out of business. What can I do myself, right now, to get the ball rolling?"

Many organizations are like aircraft carriers—they are designed to move large amounts of people and things in a given direction, usually at a steady pace. A quality conscious organization requires a different design, one that is flexible, rapid, and team-oriented. As a manager, it doesn't help you to know that the carrier is on the way when your customers demand decisive action now.

This book provides you with the ammunition you need to hold off your competitors and launch your own guerrilla attacks.

Summarizing, then, this book:

- Presents a list of actions that you personally can take immediately regardless of where other parts of the business are in their quality journey.

- Presents an overall framework and set of tools to help you make quality happen.

- Helps you connect your own individual actions to corporate quality objectives.

- Increases the level of personal satisfaction you will experience.

CHAPTER ONE

THREE STAGES OF IMPLEMENTATION

EXECUTIVE OVERVIEW

- It takes a galvanizing event to become committed to improvement.

- Once committed, companies progress through three distinct stages of implementation.

- Actions taken to improve quality are most effective when they occur at the right time (they match the correct stage for the organization at a given point in time).

WHEN DO COMPANIES COMMIT TO QUALITY IMPROVEMENT?

What causes a company to commit to quality improvement? Does the CEO come in to work and say, "It's time to improve our quality?" Similarly, what causes a person to decide to improve their level of fitness? Do they simply wake up one day and say, "It's time to get fit?"

Studies show that companies *successfully* commit to quality improvement only when a significant event occurs. (We call it a galvanizing event—one that reveals there is a problem. It could be the realization that business as usual will only accelerate loss of market share, as in the famous case of Xerox when it discovered its Japanese subsidiary was selling copiers for less that what it cost its U.S. counterparts to manufacture.)

We're not sure about what causes individuals to commit to improving fitness, but we suspect for many it's also some type of galvanizing event. For

1

example, you go to the doctor and learn that if you don't change your ways you won't have to worry about next year's annual checkup! For others, it may be that your favorite pair of pants no longer fit, or that you find yourself out of breath after a short walk on your vacation.

Once a company decides to improve quality, then what happens? For all too many, not much. A statement we once read sums it up: "When all is said and done, more is often *said* than done."

The same holds true for individuals and their fitness improvement. Every January, health clubs and diet programs are flooded with eager participants. Most are dropouts by summer.

When you stop to think about it, it seems obvious that there is no quick fix. Individuals do not become fit nor do companies improve their quality overnight. After all, how long did it take for your company to get in its present shape?

THREE STAGES OF IMPLEMENTATION

Some companies do dramatically improve their quality and many individuals become fit for life. We believe that in order for quality improvement to be effective for companies, or fitness improvement to be effective for individuals, both must progress through three distinct stages of implementation. Table 1.1 summarizes and defines each stage.

The length of time it takes to progress through each stage varies according to each individual or company. We are often asked, "How long will it take for our company to improve quality?" Our best answer is, "The same amount of time it would take for everyone in the company to improve their level of individual fitness."

Now that you've seen the three stages of implementation, some of you may be thinking to yourself, "Great, so what?" "I'm not trying to improve my fitness, but I do want to help improve quality in my part of the company." "What does all this mean to me?"

The answer is deceptively simple. By recognizing which stage (of the three) best describes the current state of your organization, department, or team, you can better choose the actions you should take to successfully improve quality *at that point in time.*

The idea that there is a "right time" for taking actions to improve quality is a key point that we feel is not well addressed in the quality literature. It is our belief that this principle of timeliness is a key determinant to successful implementation.

Here's an example that illustrates the principle of timeliness. Suppose you've made a decision to improve your fitness. You know that, in general, if

TABLE 1.1 Three Stages of Implementation

Stage/Definition	Fitness	Quality Improvement
Awareness & Assessment Organization or individual consciously recognizes a gap between where they are and where they wish to be; they are also collecting and using relevant data to quantify the gap so they can determine when it is closed.	• What shape am I currently in? What shape do I want to be in? (As measured by percent body fat, cholesterol level, pulse rate, etc.) • How do I get there? (What strategies will I use to reach my fitness objectives?) • What specific exercises and foods are best for me? (In light of my fitness objectives, which type of aerobic exercise, that is, running, biking, etc., is best for me? Should I eat foods with no more than 20% fat?)	• What is the current state of my organization, department, or process? • What is the desired state of my organization, department, or process? (As measured by customer satisfaction data, cycle time, defects, etc.) • How do I get there? (What strategies will I use to reach my quality improvement objectives?) • What specific actions should I take to get started? (In light of my quality improvement objectives, which work process should I improve, which tools and methods should I use to improve that work process?)

3

TABLE 1.1   Three Stages of Implementation   (continued)

Stage/Definition	Fitness	Quality Improvement
Integration Organization or individual incorporates actions to close the gap into their daily routine; they also build in ways to measure their progress.	• What adjustments do I need to make in my daily routine? (How do I free up the time to exercise?) • What support do I need? (Do I need to join a health club or work out with a friend or personal trainer?) • How do I know I am making progress? (Do I keep a daily exercise log, count calories, weigh and record my daily weight, etc?)	• What adjustments do I need to make in my daily routine? (How do I free up the time to devote to quality improvement?) • What support do I need? (Do I need training? How do I remove those barriers that are preventing me from currently improving quality?) • How do I know I am making progress? (Do I keep a daily quality activity log, count defects, measure cycle time, customer satisfaction, etc?)
Maintenance Organization or individual has developed habits that continuously reinforce the daily actions they take to sustain quality improvement or individual fitness; both have now become second nature.	• How do I hold the gains? (For my specific fitness objectives, what combination of exercise and diet will it take to maintain the achievements I've made? What indicators will I use to check that I am staying on track?) • How do I make fitness a way of life? (How can I celebrate my success? Reward myself for staying the course over time?)	• How do I hold the gains? (For my specific quality improvement objectives, what changes must we make to the process to maintain the desired level of customer satisfaction and process performance? What indicators will I use to check that the process is staying on track?) • How do I make quality improvement a way of life? (How can I celebrate our success? Recognize and reward everyone for staying the course over time?)

4

you burn lots of calories and don't eat much, you'll start to lose weight. Does this mean that for the first five days of your fitness improvement effort you should limit your calorie intake to 1,000 a day and run five miles a day?

For most of us, the answer is a resounding no! Why? Because the first week of improvement is too soon to take these proposed actions.

Here's another example. The top management of your company has announced that, "Continuous improvement must be adopted by all parts of the organization." You are the first in your area to complete three days of training to learn the "language" of total quality, the importance of "empowerment," and the seven quality control tools typically used for continuous improvement. On your first day back in the office, you write a memo explaining that "Quality is everyone's responsibility," and effective immediately, all associates will be accountable for achieving a 10 percent increase in customer satisfaction with up to half of their next performance review based on attaining this goal. Is this an appropriate action?

Not yet. This is an example of good intentions but bad timing. Once your group has made steady progress in learning the skills and processes of quality improvement, collected and digested customer expectations and satisfaction data, and determined and implemented the work process changes necessary to satisfy customers, then it may make sense to hold them accountable for those things that they can control.

So, how do you know when the timing is right? By using the information contained in Table 1.2 to determine which stage of implementation best describes your organization as it currently operates.

Here's how Table 1.2 works. The three stages of implementation are listed across the top. Down the left-hand side, we've listed a number of characteristics of quality-conscious organizations. Collectively, these characteristics should provide you with a snapshot of your organization's quality progress at any given point in time. Think of each characteristic as a continuum of quality progress. As your organization improves its quality, it moves through each of the three stages. The cells of the table describe a single characteristic of the organization at three different points along the quality progress continuum. In other words, the table helps you answer the question, "Based on these characteristics, what stage of progress best describes my organization right now?"

Consider the following organizations. What stage best describes each?

ORGANIZATION #1

Jim Smith is the plant manager of a chemical manufacturer with operations throughout the southeastern United States. The Lafayette Plant, where Jim works, is in the midst of a monthly operations review meeting. The Quality Steering Committee of the plant is reviewing

TABLE 1.2   Characteristics of Quality-Conscious Organizations

Stage/Characteristic	Awareness & Assessment	Integration	Maintenance
Scope of Quality Management System Deployment	• Less than 20 percent of total company is involved. • Localized initiatives may be taking place in manufacturing or operations; virtually no activity in service/sales/administrative functions. • Overall effort driven by program directives.	• Twenty to 80 percent of total company is involved. • Improvement initiatives taking place throughout the business; widespread activity in service/sales/administrative functions.	• No less than 80 percent of total company is involved. • Self-sustaining.
Process Mindset	• A few core processes have been identified, but no single individual (process owner) is accountable for end-to-end core process performance. • Internal and external supplier-customer relationships for most processes have not been examined. • Sources of value within and between processes not well understood.	• Process owners for all core processes have been assigned. • Core process health diagnosed, improved, and monitored. • Internal and external supplier-customer relationships for most processes are well-understood. • All work now viewed in process terms.	• Process owners for all core and some major support processes have been assigned. • Support process health diagnosed, improved, and monitored. • Process ownership takes hold; organization is structured around processes, not functions. • Managers focus less on functions and more on improving the hand-offs between processes.

Stage/Characteristic	Awareness & Assessment	Integration	Maintenance
Quality Infrastructure	• Corporate quality definition, policy statement, and improvement objectives exist. • Informal documentation of processes and procedures exists. • Executive level Quality Steering Committee in place to manage implementation. • Approved continuous improvement process exists. • Formal quality training exists (foundation skills).	• Corporate quality implementation plan exists. • Formal documentation of processes and procedures through ISO. • Executive level Quality Steering Committee directs process improvement activities. • Formal quality training expanded to include core process owners and senior executives. • Structures in place to facilitate exchange of quality data, best practices, benchmarking, teams, etc.	• Quality plan and business plan become one common document. • Executive level Quality Steering Committee becomes fully integrated into ongoing operations; emphasis is on sustaining quality gains. • Quality message integrated into all other training (e.g., management development, sales, product).
Data	• Financial data (lagging indicators) used to drive business decisions. • Few explicit links between process performance, business performance, and customer satisfaction exist.	• Quality data such as customer satisfaction levels, defects or cycle time data is available for decision making and improvement activities. • Quality measures.	• Quality data (leading indicators) such as customer satisfaction levels, defects or cycle time data are prime sources of data for business performance measurement. • Financial data (lagging indicators) are now aligned with and support quality system (e.g., activity-based costing).

TABLE 1.2 Characteristics of Quality-Conscious Organizations (continued)

Stage/Characteristic	Awareness & Assessment	Integration	Maintenance
Rewards/ Compensation	• Existing culture, measures or rewards do not reinforce continuous improvement actions.	• Compensation systems shifting to reflect quality goals, especially executive compensation/incentive pay.	• Structured reward/ recognition system driven by quality goals for all levels of organization, both individual and team.
Measures	• Internally focused. • Not derived from or related to specific customer expectations or sources of value. • Detection based (rather than preventive; e.g., sort good from bad).	• External focus determines make-up of measurement system. • Customer expectations and sources of value become part of measurement system.	• Predictive, prevention-oriented measurement system. • Overall precision of measures increases.
Communication	• Some formal communication on quality. • Little informal communication on quality. • Infrequently mentions quality. • Inconsistent quality message.	• A few key quality themes emerge and are transmitted consistently, (e.g., voice of the customer should be used to make process improvement decisions). • High amount of quality communication through informal channels. • New channels to support quality established, such as on-line bulletin boards.	• Formal and informal communication frequently mentions quality, customer satisfaction, and all seven quality management system elements. • Extremely consistent quality message. • Hard to distinguish quality from other business performance messages.

8

Stage/Characteristic	Awareness & Assessment	Integration	Maintenance
Prevailing Paradigms	• Today's profit and market share drives tomorrow's performance. • Each function of the business should maximize its own resource utilization. • Increasing quality increases costs and decreases productivity. • Shoot the messenger. • Quality is just another flavor of the month.	• Improvement always possible and desirable. • Don't shoot the messenger. • Trust the improvement process. • Managing the white space between processes is vital to overall improvement.	• Today's delighted customers drive tomorrow's profit and market-share performance. • The business is an integrated system; resource utilization should be optimized throughout the system even if it is "inefficient" in a given function (sub system). • Increasing quality decreases costs and increases productivity.
Commitment	• Formal commitment of company by CEO to improve quality has occurred. • Permission to proceed given. • Resources identified. • Much of the organization is skeptical. • Level of CEO commitment may exceed readiness of rest of the organization.	• CEO visits customers. • Organization's leaders visibly support the improvement process. • Executives teach quality to peers. • If top management stays the course, middle management buys in.	• CEO participates on executive level improvement teams. • CEO visits end users.

TABLE 1.2 Characteristics of Quality-Conscious Organizations (continued)

Stage/Characteristic	Awareness & Assessment	Integration	Maintenance
Customer Knowledge	• Limited. • Expectations & perceptions unknown, assumed, or not well understood. • Operational contribution to customer's perception of value implicit, unmeasured, and unmanaged. • Customer facing functions not integrated with each other or other parts of the business, especially regarding sharing of common information.	• Widespread knowledge of customer expectations. • Systems in place to collect and distribute customer data. • Individuals assigned and accountable for obtaining customer data for each core process.	• Partnerships with customers in place to develop deep understanding of customer needs. • Common stream of customer data available on demand to all functions. • Each person understands his or her role in customer satisfaction.
Improvement Process	• Sporadic. • Seen as distinct from "business as usual." • Must be mandated or carefully nurtured to grow. • Historically trial and error; focuses on treating symptoms rather than eliminating root causes.	• Quality tools and improvement process applied to quality problems. • Systematic approach to dealing with quality problems driven through Quality Steering Committee. • Customer data used to identify improvement opportunities. • Root cause analysis conducted leading to deeper understanding of processes.	• Continual. • Seen as inseparable from "business as usual." • Spontaneous. • Driven by deep understanding of root causes and type of variation.

Stage/Characteristic	Awareness & Assessment	Integration	Maintenance
Variation	• Impact of variation on goal-setting and decision-making largely unrecognized and widely misunderstood, especially outside of manufacturing. • Type of variation (i.e., common or special, present in each process largely unknown).	• Use of statistical tools to identify and eliminate sources of variation (e.g., control charts, histograms). • Management asks for statistical data on processes.	• Impact of variation on goal-setting and decision-making widely recognized, well-understood, throughout the business. • Continuous real-time monitoring of all core processes to determine type of variation present. • Improvement strategy matches type of variation present.

11

monthly improvement project progress reports, which are being presented by the improvement project teams. The teams are made up exclusively of manufacturing personnel. To date, there have been no improvement projects outside of manufacturing. Each team is following the same continuous improvement process, though each project is different. The projects were selected based on defect and cycle time reduction objectives.

ORGANIZATION #2

Jefferson Bank and Trust is a large midwestern bank that caters primarily to wealthy individuals and closely held corporations. Recently, the CEO of the bank concluded that declining profit margins were here to stay unless the bank improved its operations and focused on meeting the special needs of its customer base. To that end, she challenged the senior vice-presidents to join her in bringing total quality management to every corner of the bank. She also established three improvement objectives: reduce cycle time of the banks loan origination processes by 50 percent within six months while reducing the default rate 25 percent; achieve a level of customer satisfaction of 98 percent from the bank's top 100 customers within two years; improve gross profit margins by 25 percent within 12 months.

ORGANIZATION #3

Fashion Express is a specialty retailer of women's apparel. It is the market-share leader and the envy of the competition within the industry.

Everyone who works for Fashion Express receives stock in the company. There are no commission positions, but each employee/owner has the opportunity to earn up to 50 percent of their total compensation based on store customer satisfaction, retention, and profit objectives. Employee/owners are members of at least one "business" team. Each team is responsible for a key work process at the store. All team members have been trained in process improvement techniques and are actively working on at least one improvement project as part of their routine responsibilities. Ask anyone who works at a Fashion Express store and they can tell you who their customers are, what those customers' most important expectations consist of, and how his or her job impacts those customers. Each employer/owner has access to all store data, including financial, operations, and customer satisfaction data. In fact, there are store and team report cards, which are updated daily, that provide results on key indicators that affect customer retention, process performance, and financial performance.

Turnover at Fashion Express is virtually nonexistent, as compared with 20 percent for other retailers in the same industry. Whenever a job has to be filled, there are always hundreds of applicants for a single position. In addition to quarterly surveys of Fashion Express customers, the company surveys each employee/owner quarterly as well. The level of internal satisfaction is at least 98 percent and is even higher for those who have been with the company longer. Fashion Express has been mentioned on numerous "best places to work" lists.

How did you do? Check your understanding based on the following discussion.

ORGANIZATION #1

Stage 1. The Lafayette Plant has some quality infrastructure in place (the Quality Steering Committee), is using a consistent improvement process, and is monitoring project progress. However, the scope of its quality management system deployment is limited to manufacturing personnel. This makes the Lafayette Plant at Stage 1.

ORGANIZATION #2

Stage 1. Jefferson Bank & Trust is in the beginning of its quality improvement journey. The CEO has committed to improvement and has established specific company-wide improvement objectives.

ORGANIZATION #3

Stage 3. Fashion Express is an organization that has made considerable progress in its quality journey. It has organized its store operations around processes (business teams). Compensation for everyone is based to a large extent on meeting customer satisfaction, retention, and profit objectives. Everyone routinely works on improvement projects. Store data is available to everyone, and report cards are widely in use.

SUMMARY

- It takes a galvanizing event for organizations or individuals to successfully commit to improvement.

- Organizations and individuals move through three distinct stages as they progress along their improvement journey.

- During Stage 1 (Awareness & Assessment), organizations or individuals consciously recognize a gap between where they are and where they wish to be; they also collect and use relevant data to quantify the gap so they can determine when it is closed.

- During Stage 2 (Integration), organizations or individuals incorporate actions into their daily routine to close the gap; they also build in ways to measure their progress.

- During Stage 3 (Maintenance), organizations or individuals develop habits that continually reinforce the daily actions they take to sustain quality improvement or individual fitness; improvement is now second nature.

- The principle of timeliness suggests that there is a "right time" for improvement actions.

- By recognizing which stage best characterizes your organization, department, or team *right now,* you can make the principle of timeliness work for you instead of against you.

CHAPTER TWO

WHAT DO SUCCESSFUL MANAGERS DO, ANYWAY?

EXECUTIVE OVERVIEW

Have you ever wondered what actions managers take in companies that have successfully implemented quality improvement? Our analysis shows that effective managers take actions that are consistent with three broad principles:

- They integrate the quality message into their daily communication.

- They visibly spend time daily taking actions to improve quality.

- They align the balance of consequences to reinforce quality improvement at every opportunity.

INTEGRATE THE QUALITY MESSAGE INTO YOUR COMMUNICATION DAILY

When effective managers communicate regarding quality improvement, they use a combination of verbal, nonverbal, or written forms of communication to transmit well-understood messages to those around them. They focus on three characteristics of their quality messages.

First, they focus on the *content* of the message. Effective managers never seem to miss a chance to talk quality. The importance of meeting or exceeding customer requirements is pervasive in their day-to-day communication. Equally important is the relevance of quality improvement that these managers help convey to their associates, peers, and superiors. These managers are able to use language that is meaningful to each individual they work with, rather than simply reciting a company phrase or quoting the company quality policy to all. When people hear an effective manager talk quality, they know that this manager takes the subject seriously, has invested time to understand the issues, and sincerely believes that quality improvement will make everyone's job a little more satisfying.

The second characteristic of quality messages that effective managers attend to is *consistency.* Simply put, the message is the same, day in, day out. The *message also matches the actions* these managers take. Nothing sends a clearer signal of your real intent than this aspect of consistency. Effective managers of quality improvement walk the talk.

Finally, effective managers focus on the *frequency* of the quality messages they transmit to others. As a percentage of their total communication, effective managers talk quality more often than their ineffective counterparts.

Lets summarize. One principle guiding the actions of effective managers when implementing quality is that of integrating the quality message into their daily communication. Effective managers focus on the content, consistency, and frequency of the quality messages they send, regardless of the form the communication takes. They tend to send the right messages day in, day out, and they talk quality often. And most important, their actions match their words.

VISIBLY SPEND TIME IMPROVING QUALITY

Closely related to what managers say about quality is the *amount of time* they spend taking steps to improve quality. Visibly spending time on quality improvement is the second principle effective managers follow to implement quality.

By spending time we mean allocating time on a daily basis to specific actions that will increase customer satisfaction.

Believing that quality improvement is vital, or extolling the virtues of quality improvement, no matter how eloquently, will not in and of themselves improve quality. Only by changing the amount of time we actually spend improving the work processes we manage can we expect to make a difference in the amount of satisfaction our customers experience.

Effective managers have discovered that although they must allocate a greater percentage of time daily to quality improvement actions, it is not so much the amount of time they spend but *what they spend it on* that causes high levels of customer satisfaction.

Returning to our fitness analogy helps illustrate this principle. Suppose your fitness improvement plan calls for spending more time on exercise. You decide to increase the time you spend from 15 minutes to 30 minutes three times a week. You track your progress and find that although you have indeed doubled the amount of time you spend on exercise, your level of fitness is increasing more slowly than you had hoped. Upon closer investigation, you learn that although you are spending more time working out, the exercise you've selected (walking) does not burn as many calories in the same time period as running or using the stair climber or cross-country ski machines do.

As busy as all of us are, wouldn't it be nice to have a list of the high-payoff actions managers could take to implement a quality management system (QMS)? That way, as you plan and allocate your time, you could choose those actions that have a greater relative impact.

MANAGER ACTIONS CHECKLIST

You're in luck. We've created just such a list. We call it the Manager Actions Checklist. Think of it as a crib sheet for identifying actions you can take to help successfully implement each of the seven elements of a QMS. Each checklist contains actions that are consistent with the three principles of effective quality managers; additionally, the actions are grouped according to the three implementation stages discussed in the last chapter as well. The checklists appear at the end of each chapter in Section 2 of this book.

We've also developed a Personal Quality Planner (i.e., Quality Management System Planning Worksheet). This is a weekly time management tool designed to help you integrate actions from the seven QMS elements into your daily work. (This tool is described more fully in Chapter 11.) Together, the Manager Action Checklists and Personal Quality Planner can help you spend your time on specific actions that will make quality work in your organization.

So far we've discussed two of the three principles guiding the actions of effective managers in implementing a QMS. The first principle was integrating the quality message into daily communication, and the second was visibly spending time daily taking some action to improve quality.

ALIGN THE BALANCE OF CONSEQUENCES TO REINFORCE QUALITY IMPROVEMENT

The third and final principle is incredibly powerful, especially when integrated with the other two. Rather than accepting the status quo, effective managers align the balance of consequences their associates experience so that those consequences act to reinforce, rather than discourage, quality improvement.

Consequences are events that follow behaviors and change the probability that those same behaviors will recur in the future. Here's a quick example. You put the correct change in a soft drink machine and press the button for the drink you desire (a behavior or action). If all is well, your soft drink comes out (the consequence). But what happens if the machine takes your money (a different consequence)? Will you try again? Go to a different machine? When the consequences change, behavior changes. An extremely important set of consequences comes from what you as a manager do, say, provide, or withhold as perceived (experienced) by your associates.

We'll explain this principle more fully in Chapter 10; we'll also provide you with a tool called the Balance of Consequences Worksheet to help you make use of this powerful management technique. For now, here are a few key points to remember regarding consequences:

- Every behavior has a consequence.

- Consequences affect performance whether or not they are managed.

- Every individual experiences consequences differently; different people often respond to the same consequence differently.

- There are two basic types of consequences: those that increase performance (positive) and those that decrease performance (negative).

With the above points in mind, by balance of consequences we mean the cumulative effect of the positive and negative consequences that a person experiences after a given behavior or action.

Effective managers have learned how to analyze and align the balance of consequences so that they more strongly encourage or reinforce quality improvement.

In Chapter 1, we defined each of the three stages organizations progress through when they successfully implement a QMS. In this chapter, we defined three principles that managers follow in those companies that have successfully implemented quality improvement.

Let's see what you as a manager might do to apply these three principles at each of the three stages of implementation. Table 2.1 lists actions that are consistent with each principle as your organization progresses through each stage of implementation.

If you focus on these three principles, you will achieve more success than most managers. Chapter 11 describes a set of tools that will help you apply each of these three principles. These tools are the Personal Quality Planner, Personal Quality Checksheet, and the Balance of Consequences Worksheet.

The examples that follow illustrate how managers at different organizations apply one or more of the three principles discussed in this chapter. As

TABLE 2.1   The Three Key Principles of Effective Quality Managers

Stage/Managerial Principle	Awareness & Assessment	Integration	Maintenance
Integrate the Quality Message Into Daily Communication	• Reflect on your communication for a single day; analyze the message you've been sending regarding quality. • Monitor how frequently you mention quality in your daily communication. Monitor the consistency of the quality messages you send. • Consciously define the quality message that you want to send. • Identify both formal and informal opportunities to communicate the quality message.	• Reflect on your communication for a single day; analyze the message you've been sending regarding quality. • Broaden the focus of your quality message to include all seven quality management system elements. Tailor your messages to each recipient to demonstrate relevance to their job/role. • Continue to monitor the frequency and consistency of the quality messages you send; consider getting others to collect data to help you with this analysis. • Ask questions that prompt your associates to demonstrate how a proposed course of action improves quality.	• Reflect on your communication for a single day; analyze the message you've been sending regarding quality. • Examine the balance and distribution of your quality message communication in relation to all seven quality management system elements; is the emphasis where you want it to be? • Monitor the frequency and consistency of the quality messages that your associates send; use of quality language should be widespread and natural at this stage in meetings.

TABLE 2.1 The Three Key Principles of Effective Quality Managers (continued)

Stage/Managerial Principle	Awareness & Assessment	Integration	Maintenance
Visibly Spend Time Daily on Actions that Improve Quality	• Reflect on your actions for a single day; calculate the percent of total time spent on quality. • Analyze the time you currently spend on quality; are you spending time on detection of, or reaction to, poor quality or on those actions that will help implement the seven elements of a quality management system? • Decide how you will visibly demonstrate that you are spending time on quality improvement; plan how you will ensure that your associates see that you take quality improvement seriously. • Plan the specific actions that you will take to begin implementing each of the elements of your quality management system; start small, use the Manager Action Checklist(s), Personal Quality Planner, and Personal Quality Checksheet to help you select and monitor the actions you will take (see Chapter 11 for information on how to use these tools).	• Reflect on your actions for a single day; calculate the percent of total time spent on quality; are you allocating more time for improvement or quality management system related actions? Are the actions visible to those around you? • Determine the root causes of current process performance for each of your key work processes. • Broaden the focus of your actions to establish or strengthen all seven elements of a quality management system for each of your key work processes.	• Reflect on your actions for a single day; calculate the percent of total time spent on quality; at this stage, most of your time should be spent using the data generated by your quality management system to manage the work processes in your area, and on actions that sustain the quality management system.

20

Stage/Managerial Principle	Awareness & Assessment	Integration	Maintenance
Align the Balance of Consequences to Reinforce Quality Improvement	• Identify the existing consequences for each associate. • Determine whether each consequence is positive or negative as perceived by the associate. (In most firms, the *existing* consequences for quality improvement are perceived as *negative* by associates, that is, improvement projects add to the workload, often result in unpaid overtime or lost "work" time, often require people to work in unfamiliar teams, use unfamiliar skills, and take "unnecessary" risks by exposing "bad news" about current work processes. If you can't counteract these consequences and add some strong positive consequences, you'll know why so many firms talk quality but are unable to make real progress.)	• Establish strong positive consequences for each associate working on an improvement project. • Reduce the impact of existing negative consequences. • Look for ways to make continual improvement self-sustaining (i.e., build positive consequences for improving quality into each associates job). • Revise or establish reward, measurement, and recognition systems to reinforce continual improvement.	• Monitor the balance of consequences to ensure that continual improvement is reinforced.

you read each example, ask yourself these questions: What stage of implementation best characterizes this organization? Which principle(s) is this manager demonstrating?

EXAMPLE 1

Jerry is a manager of a regional communications center in a large transportation company. Historically, this firm has focused on rail transportation but recently has recognized the need to "reinvent" itself as a full-service intermodal provider of freight nationwide. As a result, the firm is initiating process improvement teams, establishing company-wide cycle-time-reduction improvement objectives, and "rediscovering" its customer base. Jerry understands the importance of the communications center as a means of providing customers with timely, accurate, information on their shipments 24 hours a day. However, this is a significant shift in focus for many of the people who work in the center. To help people understand this new emphasis, Jerry decides to develop a personal communication plan to help him explain what all this "quality stuff" means to the people who work in the center. The plan addresses the following:

- The current vs. the desired state of the communications center, emphasizing the major steps necessary to move to the desired state.
- Specifics on how the organization is committed to quality improvement and what support it is providing to help the center increase customer satisfaction.
- Which jobs and responsibilities will be affected, what will be different, and what will remain the same.
- How the basic concepts of quality improvement will be translated into center operations (i.e., what it means to each job in the center).

EXAMPLE 2

Cathy manages new-product development for the midwestern region of Chiptek, a manufacturer of specialized integrated circuits used primarily by the automotive industry. Even though Chiptek has a reputation for technical excellence, last year they began company-wide continuous improvement. Cathy's group is made up of highly educated professionals who tend to view themselves more as artists than scientists or engineers. When Chiptek first introduced its improvement process, the people in Cathy's area were underwhelmed, to put it mildly. Typical of the comments she sometimes heard were, "Why bother us with quality improvement, it's the manufacturing plants where we should be focusing first." Or, "If we had more time to design and test our circuits, we could improve the yield."

Fortunately for Chiptek, Cathy firmly believed that quality was everyone's responsibility, so she set out to try to better understand how each of her designers might respond personally to the company's desire to improve quality. She reasoned that if she could identify the type of experiences that would be personally motivating for each of her designers, perhaps she could create opportunities to link those experiences to improvement project participation. This way, when someone worked on an improvement project, they would perceive a direct benefit for doing so. At the same time, she wanted to reduce the obstacles or barriers that each designer might encounter while working on an improvement project. She soon discovered that for some of her designers, the opportunity to present findings at national conferences was appealing, while for others public speaking was a scary proposition. She found that some designers seemed happiest working as lone wolves, whereas others wanted the opportunity to collaborate with peers. Cathy learned that some members of her group were particularly interested in state-of-the-art breakthroughs and others liked the intellectual challenge of working on a problem that had traditionally stumped the company's best thinkers. Some liked keeping meticulous records of projects they worked on, while others found it tedious. As she reflected on her findings, Cathy thought, "It's a good thing I didn't get everyone in a room together and make improvement project assignments."

EXAMPLE 3

Richard is about to attend a progress review meeting for one of the improvement project teams he chartered. Though he is very busy, he makes it a point to attend every review meeting for each team in his area. As a regional sales manager, he could delegate some of these meetings to territory or account managers, but he is concerned about the message that would send like wildfire throughout the business. Besides, he's learned that asking good questions at these meetings helps him stay in touch with issues that are important to customers, both internal and external. Its been more than three years since his company launched its drive for continuous improvement. During that time, Richard has discovered that the more people see him do things to "move the quality yardsticks ahead," the more likely they are to do the same. He used to begin weekly sales meeting with questions regarding quarterly sales goals and profit margins. Now he asks how many customers have been visited or contacted by phone or fax, and what lessons have been learned. He considers each day a disappointment if he has not taken an action to get closer to customers or help someone else do the same.

How did you do? Check your understanding of the three key principles that effective quality managers follow based on the discussion that follows.

EXAMPLE 1

The transportation company has initiated process improvement teams, established company-wide cycle-time-reduction improvement objectives, and "rediscovered" its customer base. These are signs that the firm is early in the *first* stage of implementing a *QMS*. Jerry's development of a plan to help him better communicate with the people in the center is an example of a Stage 1 action that is consistent with the principle of integrating the quality message into daily communication. In this case, the plan helps Jerry define the quality message that he wants to send.

EXAMPLE 2

Chiptek is also in Stage 1, though probably further along than our transportation firm in Example 1. Cathy is trying to identify the existing balance of consequences for each of her integrated circuit designers. Once she understands how each of her designers is likely to experience the company's wish to improve quality, she will be better able to create a win-win opportunity for each of them. Chapter 10 explains how to analyze the balance of consequences in greater detail.

EXAMPLE 3

Richard's firm is in the second stage of implementation. In addition to manufacturing or operations, the improvement process is being used in sales. It has been three years since the company as a whole initiated its improvement process. Richard's hands-on involvement is an example of visibly spending time daily on actions that improve quality.

SUMMARY

- Managers who are most effective at quality improvement follow three key principles:

  1. They integrate the quality message into their daily communication.

  2. They visibly spend time daily taking actions to improve quality.

  3. They align the balance of consequences to reinforce quality improvement at every opportunity.

- When *effective* managers integrate the quality message into their daily communication, they focus on the content, consistency, and frequency of the quality messages they send.

- When *effective* managers spend time taking actions to improve quality daily, they identify high-leverage actions and follow the principle of timeliness, that is, they match their actions to the implementation stage (see Tables 1.1 and 1.2 in Chapter 1) that best describes *their* part of the organization.

- Finally, when *effective* managers align the balance of consequences to reinforce quality improvement at every opportunity, they identify and take the actions necessary to consciously stack the deck in favor of quality improvement.

CHAPTER THREE

WHAT IS A QUALITY MANAGEMENT SYSTEM (QMS)?

EXECUTIVE OVERVIEW

- Initially, many managers seem puzzled, confused, or skeptical when they first discuss QMS. Typical comments were:

  — "What does this have to do with us?"
  — "I'm not sure what you mean by a 'management system.' "
  — "Isn't this the responsibility of our senior executives?"
  — "The only reason we're doing this is because someone high up wants us to apply for the Baldrige award."

- This chapter introduces the seven elements that make up a Baldrige type QMS:

  — Total Customer Satisfaction
  — Quality Results
  — Strategic Quality Planning
  — Process Quality Assurance
  — Management by Fact
  — Human Resources Effectiveness
  — Leadership

- There are several books on the Baldrige Award. This book is not intended to replace those texts. Instead, the purpose of this book is to help managers understand what they personally can do to help implement each management system element in the work processes they manage.

WHAT IS A QMS?

Let's start off by defining what we mean by a QMS. Simply put, a QMS is a set of values that describes how to run the business as a whole. Collectively, the values reflected in the processes that make up a QMS tell us what to focus on or emphasize as we manage a given work process.

Perhaps one reason for the confusion some managers experience is that the words "quality," "management," and "system" have not traditionally been used in the context of describing a business philosophy.

For example, many managers associate the word "quality" primarily with product characteristics, and as a result believe that quality is manufacturing or operation's responsibility.

The word "management" is more familiar, but interestingly enough, always seems to describe the next higher level of authority in the organization.

Finally, the word "system" conjures up various meanings—from specific information or computer systems to more general intents such as "you can't beat the system" or "you have to go around the system to get any real work done."

With the preceding frame(s) of reference, it's no wonder that the term quality management system causes glassy eyes, blank stares, and furrowed brows.

Here is the frame of reference we have when we use the term quality management system.

FIGURE 3.1   Common Reactions When Discussing Quality Management Systems

A system is a group of related processes, and a process is a set of re-peatable activities that transform inputs into outputs that have value to inter-nal or external customers. So a system, then, is a *collection* of processes whose purpose is to create value for customers. See Figure 3.2.

Keeping the idea of a system in mind, a QMS is a special type of system (a particular group of processes) whose purpose is to *manage* quality, the key distinction being that by quality we mean total customer satisfaction rather than just product characteristics.

Now when you hear or read the words quality management system you'll know they really mean, "a particular group of processes whose purpose is to produce satisfied customers." See Figure 3.3.

If you accept the preceding definition, then as a manager interested in achieving customer satisfaction, two questions should come to mind: "What are the related processes that are intended to produce satisfied customers?" and "What do those processes mean to me?" (What are my responsibilities? What actions can I take to successfully perform those processes? How does each process relate to (impact) how I manage a given work process?)

The remainder of this chapter introduces each of the processes that make up a QMS and lists the relevant concerns for each. Chapters 4 through 10 further describe each process.

Figure 3.4 represents a typical work process, consisting of inputs being transformed into outputs by a series of repeatable steps.

Each process that makes up a QMS has a specific effect on any work process. To effectively implement the processes that make up a QMS you must understand how each process links to, impacts, or interfaces with a work process.

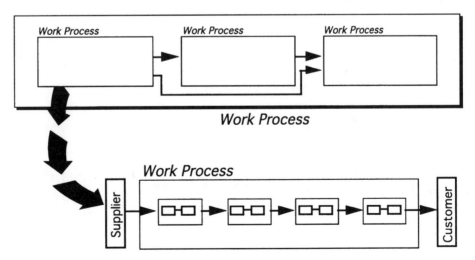

FIGURE 3.2   A System Consists of Related Processes

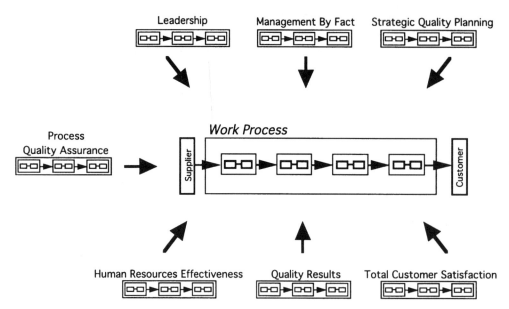

FIGURE 3.3   What is a Quality Management System?

Using the work process graphic in Figure 3.4 as a baseline, we will illustrate the link, or interface, each QMS process has and discuss its impact, one at a time, beginning with Total Customer Satisfaction.

TOTAL CUSTOMER SATISFACTION

Keep in mind that the purpose of a QMS is to manage quality—which means increasing the level of satisfaction that customers (both internal and external) experience throughout the business.

Total Customer Satisfaction consists of those processes that define internal and external customers for every work process, determine the requirements of those customers, and determine whether and to what extent those

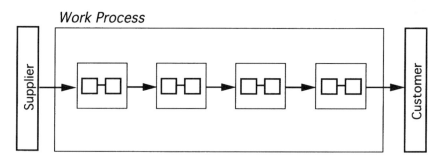

FIGURE 3.4   Typical Work Process

customers are satisfied with the outputs that the work processes of the business provide on an ongoing basis. See Figure 3.5.

As a manager, the questions you should ask yourself with regard to Total Customer Satisfaction are:

- Who are the customers for this process?

- What are their requirements?

- How well are we meeting those requirements? Based on what data?

- What action is necessary for those requirements that we are not currently meeting?

The primary outputs of Total Customer Satisfaction processes are customer expectations data and customer perception data. These outputs act as inputs to other QMS processes so that decisions regarding the operation of a work process may be made; that is, they provide feedback to continue the process as is, improve the existing process, or design a different process.

For example, there is a critical relationship between the outputs of Total Customer Satisfaction and Quality Results.

QUALITY RESULTS

Whereas Total Customer Satisfaction provides external feedback (from outside the work process), Quality Results processes provide internal feedback (data

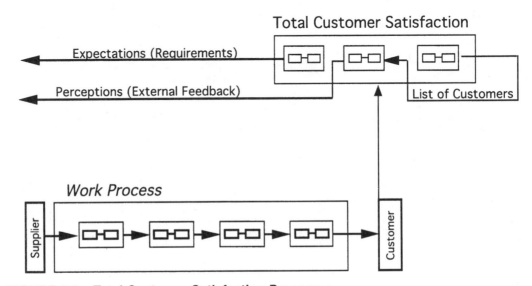

FIGURE 3.5   Total Customer Satisfaction Processes

on process performance before a customer receives the output). Done well, Quality Results processes act as a early warning system for customer satisfaction; that is, they use real-time measures to predict how customers will perceive the outputs they subsequently receive while the process is operating. See Figure 3.6.

Key Quality Results questions include:

- What aspects of process performance are we currently measuring?

- How well is the process performing?

- Do our process measures help us manage the process or are we simply tallying low-value data? (Counting scrap vs. monitoring critical variables in the process that prevents scrap.)

- What happens to the measurement data we collect? Who receives it? When do they receive it? How is it used?

Quality Results data should help us determine whether a work process is stable (variation is within a predictable range), and whether the process is consistently meeting customer requirements. Just as customer expectations data is referred to as the voice of the customer, Quality Results data represents the voice of the process. Both voices are necessary to manage work processes so that they satisfy customers.

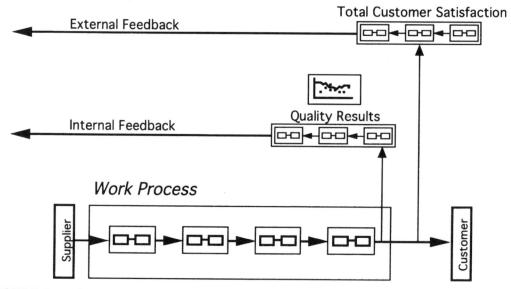

FIGURE 3.6   Quality Results Processes

The data obtained via your Total Customer Satisfaction and Quality Results processes feed into a third set of processes known as Strategic Quality Planning.

STRATEGIC QUALITY PLANNING

When we conduct workshops explaining the various elements of a QMS, participants often ask, "What does Strategic Quality Planning have to do with us? Isn't that the job of senior management?"

The short answer to the question is that Strategic Quality Planning is vital in every manager's job, because *Strategic* refers to the impact that a given process has on customer satisfaction. The greater the impact on customer satisfaction, the more strategic the process is. Stated another way, the more value a given process adds to the customer, the more strategic that process is. See Figure 3.7.

Thus, Strategic Quality Planning means developing specific plans to run your part of the business so that the quality of those processes that most impact customer satisfaction are managed, measured, monitored, and improved.

Collectively, your Strategic Quality Planning processes should help you answer these questions:

- What processes most impact customer satisfaction?

- Based on customer expectations data, what are the desired process performance targets or standards?

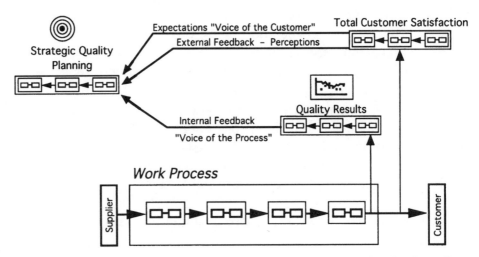

FIGURE 3.7   Relationship Among Total Customer Satisfaction, Quality Results and Strategic Quality Planning Processes

- Based on Quality Results data and customer perception data, are the processes satisfying customers?

- What are our process improvement objectives?

- What is our plan for daily management, measurement, monitoring, and improvement of our strategic processes?

Once you have identified the work processes that most impact customer satisfaction and developed a plan for managing the quality of those processes, you must execute your plan. Usually this involves continually examining and improving the performance of the work process to reduce the amount of variation present and to increase the value the process adds to the customer. The processes that help you do this are your Process Quality Assurance processes.

PROCESS QUALITY ASSURANCE

Process Quality Assurance is another term that seems to cause confusion among workshop participants. "Isn't this what the quality department does?" someone invariably asks.

Let's take a closer look at what Process Quality Assurance processes mean to you as a manager. First of all, who is responsible for the quality of the work processes in your area, you or the quality group? Put another way, if there is a "quality" problem with one of the processes that you manage, whose neck is on the line?

Process Quality Assurance means taking action to create high levels of customer satisfaction on an ongoing basis. These are the processes in which the "technical" aspects of quality improvement take place.

Process Quality Assurance includes those processes that gradually improve the performance of a work process, such as continuous improvement, *Kaizen,* or Shewhart's PDSA cycle, and those that are intended to produce quantum leaps in process performance, such as benchmarking or reengineering. They also include the definition, establishment, and deployment of the measures you use to assess and monitor how well your work processes are performing. See Figure 3.8.

Your Process Quality Assurance processes should answer these questions:

- Are we measuring the right things in light of customer expectations?

- What measures should we have in place to help predict customer satisfaction?

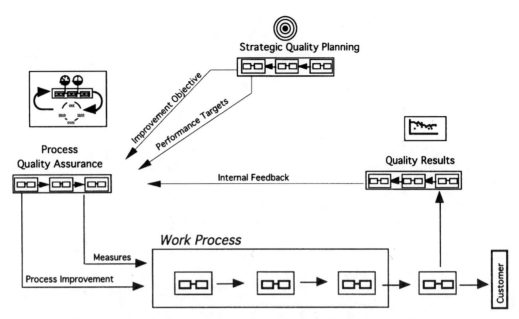

FIGURE 3.8 Relationship Among Strategic Quality Planning, Quality Results and Process Quality Assurance

- Have we identified the critical factors within our work processes to measure and manage in order to produce (cause) high levels of customer satisfaction?

- Based on our process-improvement objectives, what process-improvement projects are necessary?

- How capable are our processes at meeting customer expectations?

- Have we identified the types and sources of variation in our processes?

So far we have discussed four of the seven elements of a QMS. Figure 3.9 shows how these four elements relate to one another and serve as the foundation of customer-focused process measurement.

The next three elements, Leadership, Management by Fact, and Human Resources Effectiveness, do not relate to any work process, per se. Rather, they consist of processes *that transcend but affect all work processes.*

MANAGEMENT BY FACT

Management by Fact processes are those decision-making processes that use data to help us determine a particular course of action.

Much of a manager's job involves making decisions. This is not new. What is new is the basis or methods used to make those decisions. Two thoughts

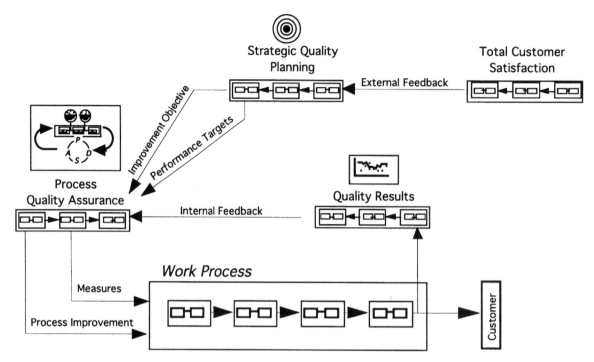

FIGURE 3.9   Relationship Among Total Customer Satisfaction, Strategic Quality Planning, Process Quality Assurance, and Quality Results Processes

are key. First, you must overcome the tendency (if it exists) to use gut feeling, hunch, tradition, hearsay, past experience, superstition, or judgment as the sole basis for your decisions. Second, you must always ask yourself, "What data is relevant to this decision, and how will I use that data to select a course of action?" Usually, you will find yourself using some combination of the seven quality control tools or another relevant set of tools, along with data to help you make effective decisions.

Management by Fact processes make use of the data provided by your Total Customer Satisfaction and Quality Results processes. This is the point at which the voice of the customer and the voice of the process come together to help you make decisions regarding the work processes in your area. See Figure 3.10.

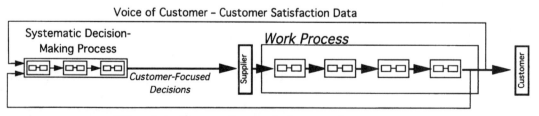

FIGURE 3.10   Management By Fact

Your Management by Fact processes provide you with the answers to these questions:

- Is increased customer satisfaction the driving force behind our decisions?

- Are we using customer expectations, perceptions, and process-performance data to make decisions regarding process performance, process improvement, and customer satisfaction?

- On what basis are performance goals, quotas, or standards being set?

- How do we know when—and when not—to change a process?

HUMAN RESOURCES EFFECTIVENESS

Another management system element that affects all work processes is Human Resources Effectiveness. It is hard to find a work process that does not depend on people to do, or help do, the work. See Figure 3.11.

Whereas Process Quality Assurance processes consist mostly of the technical side of quality improvement, Human Resources Effectiveness

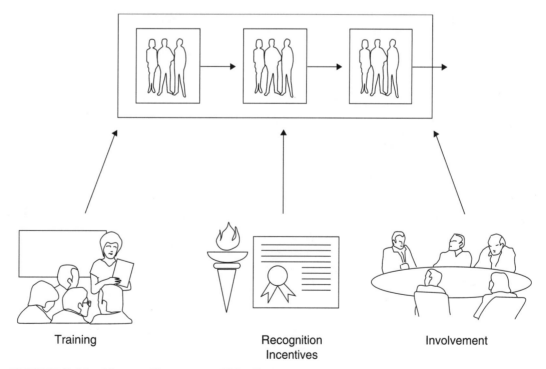

Training   Recognition Incentives   Involvement

FIGURE 3.11   Human Resources Effectiveness

processes address the people or human aspects. They are the flip sides of the quality improvement coin. See Figure 3.12.

No organization will be able to successfully implement a QMS without the full support of its people. People are the engine that drives quality improvement. People Effectiveness processes are those processes that not only sustain the engine, but help the engine run well.

As a manager, you should keep three things in mind with regard to Human Resources Effectiveness. First and foremost, in the eyes of your employees *you* are the organization's people policies and procedures as well as all the other elements of the management system. What *you* say and do sends a direct message about what the organization values, expects, and rewards. Second, the most powerful impact you can have on your employees is to align the balance of consequences the organization provides with the personal values of each employee. This is the fastest way for your employees *and* the organization to realize their potential. Third, the implementation of a QMS will require you to spend a greater percentage of your daily time training, developing, and coaching your employees.

Your Human Resources Effectiveness processes should answer these questions:

- Do our "people" policies and practices help or hinder our ability to satisfy customers?

- If people that report to me are also my customers, how satisfied are they with me as a supplier?

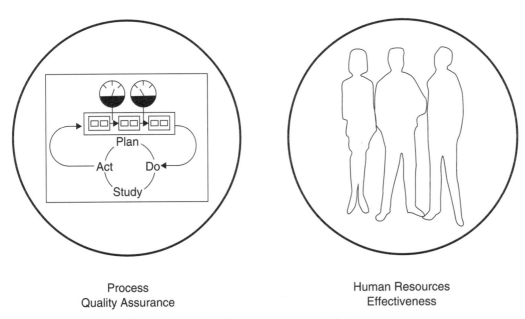

Process
Quality Assurance

Human Resources
Effectiveness

FIGURE 3.12   Two Sides of the Quality Improvement Coin

- Does the current balance of consequences that each associate experiences reinforce quality improvement? Does it support their personal values?

- How can I build-in rather than add-on training, development, and coaching opportunities for all my associates?

LEADERSHIP

The final management system element is Leadership. Though most discussions of this element usually occur first, we have chosen in this overview to cover it last—not because it is unimportant, but because we feel it makes more sense once you are familiar with the other six elements.

The reason for this is straightforward. Leadership is a set of processes intended to make sure that all the other processes that go together to comprise the QMS take place and are well integrated. Thus, Leadership not only formulates the management system but is the glue that holds it together as well.

There is only one question that your Leadership processes should address, but it is a broad one: Am I taking the right actions (to establish, institutionalize, and reinforce all the processes of a QMS) right now? See Figure 3.13.

FIGURE 3.13   Leadership

Ironically, the notion of a QMS is not new. In simpler times, what we today call a QMS probably would have been called good business practice. As you read the following scenario, try to identify which management system element or process is being applied. *Hint:* Not all seven elements may be present.

Suppose you were a craftsperson. Your name is Chuck. One of your customers, Frank, who is also a friend and neighbor, comes to you and says, "Chuck, can you make me a harness?" "Of course," you say. You proceed to ask a number of questions. "How will the harness be used? When do you need it? Do you have a particular material in mind? What size harness?"

The two of you agree on the job and you begin making the harness. As work proceeds, you keep your customer informed of progress and involve him in key decisions that data from your past harness-making jobs tells you are important to him.

Over the years, you've kept careful notes regarding your harness-making techniques. You've listened to your customers, learned what they like and dislike, and how they use the harnesses you make. As a result of your knowledge, experience, and lessons learned from customers, you know which part of your work affects the finished harness. You measure twice and cut once to prevent mistakes. You use only the best raw materials and tools. Your cutting and sewing techniques have been refined so that you get the most out of your materials, and the finished harness is reinforced at critical stress points. You've earned the respect of your customers because they know you craft a well-made harness that makes their lives a little easier.

When the harness is finished, you deliver it to your customer to make sure it fits and to ensure that he knows how to make the proper adjustments and take care of it to make it last. You ask if he has any questions and if he is satisfied. Just before you leave, you say, "Let me know how the harness works. If you don't mind, I'll check back with you in a month just to make sure everything is okay." When you return to your shop, you write on next month's calendar, "Check on Frank's harness."

Today, most managers are no longer this close to their customers. They work in organizations that tend to compartmentalize or specialize the work they do on behalf of customers. As a result, for managers to do all the things that Chuck the harness-maker did, they must consciously work to integrate all seven quality management system processes into their work processes. Stated another way, managers must reconnect or link together all the processes that the organization has separated in time and location.

Which management system elements did you pick? You can check your understanding of Chuck's use of selected management system elements by referring to the following discussion. We've <u>underlined</u> the elements present, and indicated in parentheses which element the underlined words represent.

Suppose you were a craftsperson. Your name is Chuck. One of your customers, who is also a friend and neighbor, comes to you and says, "Chuck, can you make me a harness?" "Of course," you say. <u>You proceed to ask a number of questions. "How will the harness be used? When do you need it? Do you have a particular material in mind? What size harness?"</u> (TCS)

The two of you agree on the job and you begin making the harness. As work proceeds, <u>you keep your customer informed of progress, and involve him in key decisions that data from your past harness-making jobs tells you are important to him.</u> (TCS/MBF)

Over the years, you've kept careful notes regarding your harness-making techniques. You've <u>listened to your customers, learned what they like and dislike, and how they use the harnesses you make.</u> (MBF) <u>As a result of your knowledge, experience, and lessons learned from customers, you know which part of your work affects the finished harness.</u> (SQP) <u>You measure twice and cut once to prevent mistakes. You use only the best raw materials and tools. Your cutting and sewing techniques have been refined so that you get the most out of your materials, and the finished harness is reinforced at critical stress points.</u> (PQA) You've earned the respect of your customers because they know you craft a well-made harness that makes their lives a little easier.

When the harness is finished, you deliver it to your customer to <u>make sure it fits, and to ensure that he knows how to make the proper adjustments and take care of it to make it last. You ask if he has any questions, and if he is satisfied. Just before you leave, you say, "Let me know how the harness works. If you don't mind, I'll check back with you in a month just to make sure everything okay." When you return to your shop, you write on next month's calendar, "Check on Frank's harness."</u>(TCS)

SUMMARY

The first column of Table 3.1 summarizes the seven elements of a Baldrige-type QMS. For each element, columns two to four list typical processes, one or more key themes, and representative values associated with the element. See Table 3.1.

TABLE 3.1

Management System Element	Typical Processes	Key Theme	Representative Values
Leadership	Communication. Coaching. Recognition.	Integrating/linking	Lead by example.
Management by Fact	Decision-making.	Systematic approach; scientific method.	Data should be used to make process-improvement decisions.
Total Customer Satisfaction	Customer data collection.	External data collection.	Voice of the customer should drive the business.
Quality Results	Process performance data collection.	Internal data collection.	You must measure quality before you can manage it.
Strategic Quality Planning	Business planning; budgeting.	Planning.	Emphasize the value that processes create for customers.
Process Quality Assurance	Continuous improvement.	Improvement.	Reduce variation; design processes to prevent defects.
Human Resources Effectiveness	Training; aligning the balance of consequences.	Realization of potential; empowerment; collaboration.	Lifelong learning; cooperation vs. competition.

QUALITY MANAGEMENT SYSTEM— OVERALL CHECKLIST

This checklist contains each of the questions managers will be able to answer once the processes associated with all seven management system elements are in place. The list is divided into two parts. Part I contains questions that should be answered for *each* work process you manage, preferably in the order listed. Part II contains those questions that pertain to *all work processes collectively.* The items listed are subgrouped by management system element.

Part I. For Each Work Process that You Manage:

Total Customer Satisfaction

- ❏ Who are the customers for this process?
- ❏ What are their requirements?
- ❏ How well are we meeting those requirements? Based on what data?
- ❏ What action is necessary for those requirements we are not currently meeting?

Quality Results

- ❏ What aspects of process performance are we currently measuring?
- ❏ How well is the process performing?
- ❏ Do our process measures help us manage the process, or are we simply tallying low-value data? (Counting scrap vs. monitoring critical variables in the process that prevents scrap.)
- ❏ What happens to the measurement data we collect?
- ❏ Who receives it?
- ❏ When do they receive it?
- ❏ How is it used?

Strategic Quality Planning

- ❏ What processes most impact customer satisfaction?
- ❏ Based on customer expectations data, what are the desired process performance targets or standards?
- ❏ Based on Quality Results data and customer perception data, are the processes satisfying customers?
- ❏ What are our process-improvement objectives?
- ❏ What is our plan for daily management, measurement, monitoring, and improvement of our strategic processes?

Process Quality Assurance

❑ Are we measuring the right things in light of customer expectations?

❑ What measures should we have in place to help predict customer satisfaction?

❑ Have we identified the critical factors within our work processes to measure and manage in order to produce (cause) high levels of customer satisfaction?

❑ Based on our process improvement objectives, what process-improvement projects are necessary?

❑ How capable are our processes at meeting customer expectations?

❑ Have we identified the types and sources of variation in our processes?

Part II. For All Work Processes that You Manage:

Management by Fact

❑ Is increased customer satisfaction the driving force behind our decisions?

❑ Are we using customer expectations, perceptions, and process performance data to make decisions regarding process performance, process improvement, and customer satisfaction?

❑ On what basis are performance goals, quotas, or standards being set?

❑ How do we know when—and when not—to change a (work) process?

Human Resources Effectiveness

❑ Do our people policies and practices help or hinder our ability to satisfy (internal and external) customers?

❑ If people that report to me are also my (internal) customers, how satisfied are they with me as an (internal) supplier?

❑ Does the current balance of consequences that each associate experiences reinforce quality improvement?

❑ Does it support their personal values?

Leadership

❑ Am I taking the right actions (to establish, institutionalize, and reinforce all the processes of a QMS) right now?

CHAPTER FOUR

LEADERSHIP

"Leadership is practiced not so much in words as in attitude and in actions."

—Harold Geneen

EXECUTIVE OVERVIEW

- Leadership is a process.
- Leadership is not unique to managers; anyone can be a leader for quality.
- Leadership is responsible for establishing a customer focus.

WHAT IS LEADERSHIP?

Leadership is the process of energizing commitment. Effective leaders achieve higher levels of commitment primarily through communication and actions that tap into the values of those around them.

LEADERSHIP AND MANAGERS

Leadership is not unique to managers; anyone can be a leader for quality. However, it is vital that managers demonstrate visible commitment to, and actively participate in, quality improvement activities because of their role in the organization.

Once an organization announces that it intends to improve quality, most everyone adopts an "I'll believe it when I see it" attitude. And the first place they look is at senior management in general and their immediate manager in particular.

This leads to the "fishbowl effect," wherein everything their manager says and does, and in some cases doesn't do, becomes subject to close scrutiny. Simply put, as an organization increases its quality awareness, associates take their cues directly from their managers.

As a result, effective quality leadership means integrating the values and practices of a quality management system (QMS) into your daily work as a manager, not as an afterthought or when the schedule permits it, but as your primary concern.

WHY IS IT SO IMPORTANT FOR MANAGERS TO DEMONSTRATE LEADERSHIP?

To paraphrase the famous Vince Lombardi quote, "Leadership isn't everything, it's the only thing." Leadership legitimizes the quality improvement process, allocates (makes) time and resources to improvement projects, empowers employees to do their best to satisfy their internal and external customers, and, most important, sets an example. Effective leadership is the fuel of the customer satisfaction engine.

Quality leaders do much more than say the right things. They do more than encourage others to improve quality. A quality leader *acts* to ensure that the focus of all work is on satisfying internal and external customers. Managers who are quality leaders measure their success by determining the degree to which everyone in their area has institutionalized a high level of commitment to customer satisfaction.

FIVE REASONS WHY QUALITY IMPROVEMENT EFFORTS FAIL

Why do so many quality improvement efforts fail or produce only marginal results? What actions can the effective leader take to counteract the tendency to revert to business as usual? To find the answers, we must return to our fitness analogy, which asserts that quality improvement is similar to undertaking a fitness program.

The reasons why many quality improvement efforts fail are the same as those that prevent people from staying on fitness programs or diets. In fact, adopting a company-wide continuous improvement process is like asking everyone in the company to undertake a fitness program at the same time!

Here are five key reasons why fitness programs fail. No doubt there are additional reasons you may wish to add to the list, but these seem to be sufficient to cause most people to discontinue their fitness activities.

REASONS WHY FITNESS PROGRAMS FAIL

1. Rate of change too rapid

2. Amount of change too drastic

3. Balance of consequences favors status quo

4. Level of commitment too low

5. Rate of progress seems too slow

Let's take a closer look at each of the five reasons.

Reason #1. The rate of change is too rapid. People have a comfort zone when it comes to change of any kind. When faced with an opportunity for change, some people leap at the chance, some move very cautiously, and others seem to dig-in or actively avoid the change.

Many fitness programs require people to change at a rate that is too fast for a given individual's comfort zone. This causes the experience to be negative rather than positive, so people drop out to return to their comfort zone.

The same is true of many quality improvement programs. Rather than allowing individuals to improve quality within their existing comfort zones, they are designed to be "one size fits all," when flexibility and customization are much more desirable.

Reason #2. The amount of change is too drastic. The comfort zone also applies to the amount (magnitude) of change a person perceives to be associated with an improvement opportunity. A particular change may seem like taking baby steps or child's play to one person while to another it represents climbing Mount Everest!

For example, suppose a key component of a fitness program involved changing one's diet to limit the amount of simple carbohydrates (sweets) one ate. Some people would have no trouble switching to fruits or vegetables, while those of us with firmly entrenched eating habits (no dinner is complete without dessert) would find this to be a significant challenge, to say the least.

Often, a key message that accompanies quality improvement ef-forts *seems* to be, "Forget everything you've done all your working life; here's the right way now!" This is the functional equivalent of saying, "Give up all the foods you like; from now on it's broccoli for you, prefer-ably steamed!"

Reason #3. The balance of consequences favors the status quo. Most people who participate in fitness programs find they must try to fit their exercise and eating plans into their existing lifestyle. Usually this means trying to make time to work out and eat right when your day is already filled up with things that you'd rather do anyway.

Suppose you decide to exercise three days a week for 30 minutes a day (at the proper training rate, of course). You really want to do this, so you join a health club on a trial basis and plan your exercise schedule for the upcoming week for Monday, Wednesday, and Friday. It's Monday morning, the boss comes in and says, "We've got to re-vise our budget for next year. The home office wants the new figures tomorrow. Can you work late tonight?" No problem, you say. You make a mental note to change your workout to Tuesday. It's now Tuesday. You worked late last night, but got the new budget numbers. The boss is pleased. He says, "I really appreciate the hard work you put in. I was going to present the revised budget myself, but since you did all the work, I think you should get the credit too. How would you like to present the revised budget yourself at headquarters Wednesday morning?" On the way to the airport, you make a mental note to change your workout schedule to exercising three days a week—starting next week.

Wouldn't it be a lot easier to undertake and maintain a fitness pro-gram if you could somehow insulate yourself from the influences of the outside world? First, you might eliminate all distractions so nothing would interrupt you. Everything you need would be immediately avail-able, easy to use, and fun. You would have individually designed exer-cise activities complete with personal trainers, world class facilities, and access to the best athletes in the world so you could get some pointers on your favorite sport. To top it all off, you would get paid big bucks for even the slightest progress, daily. Now that's a fitness pro-gram for the rest of us!

Consider the differences between the above scenarios. In the first scenario, the consequences favor not working out. Things keep com-ing up that prevent you from going to the health club to work out. In the second scenario, however, the consequences don't just favor working out—they practically guarantee it.

The number one reason why most fitness programs fail is that the balance of consequences does not favor the actions associated with becoming fit. The same situation is true when it comes to quality improvement. In fact, in most organizations the existing balance of consequences *strongly discourages* quality improvement. A common example of this is the time pressure people feel when they are asked to work on an improvement project *on top of everything else they have to do.*

As a result, effective leaders examine existing consequences, work to remove those that discourage quality improvement, and establish consequences that encourage quality improvement. Since the balance of consequences has such a dramatic effect on quality improvement, we have devoted a section in Chapter 10, "Human Resources Effectiveness," to explain how you can greatly increase the chances of successful quality improvement in your area.

Reason #4. The level of commitment is too low. There is a threshold that separates good intentions from purposeful actions. Effective fitness programs require active participation. It's not enough to want to be fit or to agree philosophically that lowering your cholesterol is good for you. To become fit you must *act* differently; that is, you must change one or more actions in order to reach your fitness objectives. Everyone has a unique threshold. Some people will change their behavior as soon as they realize they've gained a couple of pounds or found out that their cholesterol level is too high. For others, these same events are interpreted as nuisance factors, or facts of life, that is, events over which they have no control. For these people, no fitness program will work—they have not yet reached their threshold for action.

The same is true for quality improvement. There are people who won't be ready to change how they work simply because they haven't yet reached their *personal* threshold for action. Trying to encourage people who haven't yet reached their action threshold is like trying to teach a pig to sing: It won't work anyway, and it irritates the pig.

Reason #5. The rate of progress seems too slow. Interestingly enough, though it takes weeks and months to gain weight, when we decide that we want to lose weight, we want to do so immediately. That's why, even though studies show that crash diets don't work, millions of dollars are still being spent on them every year.

Programs that promise you will lose 30 pounds in 30 days do not fundamentally change your behavior—they don't substitute the right actions for your current actions. They also do not put the right consequences in place to hold the gains, so sooner or later, the weight comes back.

The challenge is to design a fitness program that shows you are making progress so you can readily see some results from your efforts. Otherwise, you may lose interest quicker than you lose weight!

Quality improvement is no different. Often, companies mistakenly embark upon a crash program to improve quality, or cut costs, or both. These programs do not get at the root cause(s); at best, they treat symptoms. It's no surprise that sooner or later it's business as usual—until the next crisis comes along.

Now that we've described five reasons why fitness and quality improvement programs fail, what are the implications for you as a leader?

Table 4.1 lists the recommended strategies that you can take to address each of the five key reasons why quality improvement efforts fail.

TABLE 4.1

Reasons Why Fitness/Quality Improvement Programs Fail	Recommended Strategy
Rate of Change Too Rapid	• Allow people to proceed at their own pace. • Provide coaching to help increase their comfort level with improvement processes, tools, etc.
Amount of Change Too Drastic	• Do not assign everyone to an improvement project. • Assign, or better yet, let individuals choose, improvement projects or activities whose scope seems to fit their comfort level.
Balance of Consequences Favors Status Quo	• Examine existing consequences. • Remove or minimize consequences that discourage quality improvement. • Establish or strengthen consequences that encourage quality improvement.
Level of Commitment Too Low	• Do not assign or mandate quality improvement projects to everyone. • Make participation voluntary, but visibly reward volunteers.

Reasons Why Fitness/Quality Improvement Programs Fail	Recommended Strategy
Rate of Progress Too Slow	• Provide reinforcement, recognition, and feedback at frequent intervals. • Reward progress as well as outcomes. • Reward attempts. • Break projects into smaller chunks.

LEADERSHIP AND CONSEQUENCE MANAGEMENT

One of the most significant ways to demonstrate leadership for quality is to take the actions necessary to ensure that the balance of consequences your associates face strongly encourages quality improvement activities. The information and steps you need to do this are explained in Chapter 10, "Human Resources Effectiveness".

The following examples illustrate how managers at three different organizations apply quality leadership practices. As you read each example, ask yourself these two questions: What stage of implementation characterizes this organization? and, Are the actions this manager takes appropriate?

EXAMPLE 1

Chris, a manager of a Printed Circuit Board (PCB) soldering process, must decide how to achieve this improvement goal:

Reduce defects per unit in the PCB soldering process by $\geq$ 68 percent within three months.

Chris uses Pareto analysis and learns that three defect types account for 75 percent of all defects, so he decides to sponsor three improvement teams. Each team has the charter to reduce a specific defect type and is made up of performers who have in-depth knowledge of the steps of the soldering process believed to be associated with the defect type under investigation. Then, Chris:

- Contracts with an internal quality consultant experienced in process improvement to work with all three teams.

- Revises the improvement team's workloads so they have ample time to meet weekly during normal work hours to work on their improvement projects.

To further reinforce the message that quality improvement is important all the time, Chris:

- Revises the weekly staff meeting agenda to include a review of improvement project progress.

- Prepares for and actively participates in each of four improvement project checkpoints as the teams proceed through the company's improvement process.

- Rewards and recognizes team members for meeting milestones and completing the project.

- Directs each team to keep a project diary that summarizes accomplishments, identifies factors that seem to contribute most to helping the team, and lists obstacles, pitfalls, and lessons learned through each step of the improvement process.

Chris knows that without changing the way work is performed and managed, quality improvement will become the next "flavor of the month." To overcome the natural tendency to revert to business as usual, Chris:

- Gathers information from team members, team leaders, internal quality consultants, suppliers, and customers to help determine how well the improvement process is working and how it may be strengthened.

- Reviews and analyzes the project diaries from each team during each of the checkpoints and at the conclusion of the improvement projects.

- Makes sure that performance measurement and reward systems are aligned and focused on important internal and external customer expectations.

- Appoints a process sentinel (someone with ongoing responsibility for process performance) to monitor customer satisfaction in light of the improvements made.

EXAMPLE 2

Pat is a manager in accounts payable. Pat has just finished attending an operations review meeting in which the company's plans for quality improvement were discussed.

Pat decides to keep an open mind, even though she thinks that quality improvement is better suited to the manufacturing plants rather than the "knowledge workers" in her department.

No specific improvement project or objective was assigned to Pat during the meeting. However, to be on the safe side, Pat:

- Assembles a list of resources available to help learn more about quality improvement.

- Begins reading articles that discuss how quality has been improved in white collar jobs.

- Schedules herself to attend the company's introductory training course in quality improvement.

After completing the training, Pat decides to:

- Identify the major processes that make up the work in her department.

- Identify the customers for each of those processes.

- Determine the requirements and current satisfaction levels of her department's customers.

EXAMPLE 3

According to an article in the December 1993 issue of *Training & Development,* Mr. Horst H. Schulze, president and chief operating officer of the Baldrige Award-winning hotel chain Ritz-Carlton, has established the following corporate improvement goals:

- Decrease defects in 18 key processes involving customers to no more than four defects per million encounters.

- Reduce cycle time for selected processes by 50 percent.

- Achieve 100 percent customer retention, all by 1996.

The article goes on to list a number of actions Mr. Schulze regularly takes. For example:

- At a new Ritz in Hong Kong, Mr. Schulze gave a pair of housekeepers an impromptu lesson on how to make a bed.

- When meeting new employees, he uses this introduction: "Hello, I'm Horst Schulze. I'm the president, and I'm a very important person around here. [A long pause to let that sink in.] But so are you. In fact, you are more important to customers than I am. If you don't show up, we are in trouble. If I don't show up, hardly anyone would notice."

- When new hotels are opened, Mr. Schulze personally leads the seven days of training that precede every opening; he also trains the managers of all 30 properties every year.

You may check your understanding of leadership actions by reading the following discussion:

EXAMPLE 1

If you thought Chris was taking Stage 2 actions, you were right on target. Chris applied a continuous improvement tool (Pareto analysis) himself to analyze defects. He provided resources (such as an internal quality consultant and time) to assist the teams. Chris actively participated in project checkpoints and provided recognition to team members throughout the project. He also took steps to strengthen the improvement process (via the project diaries) and institutionalize the project results by establishing a process sentinel.

EXAMPLE 2

What about Pat? Pat is early in Stage 1. She is a quality novice, but she is taking the important first steps to learn more about quality and the work processes that occur within her department.

EXAMPLE 3

Horst Schulze is an outstanding example of Stage 3 actions. Though his company, Ritz-Carlton, has already won the Malcolm Baldrige National Quality Award, he has set aggressive improvement goals. He also leads by example (showing the housekeepers how to make a bed), and is personally involved in conducting training for all new hotels. His introduction to new hires sends a powerful message about the importance of focusing on the customer.

SUMMARY

- Leadership is the process of energizing commitment.

- Anyone can be a leader for quality; however, it is vital that managers demonstrate visible commitment to QMS values and processes.

- Effective leadership focuses on the actions necessary to satisfy and retain internal and external customers.

- Effective quality leaders take action to overcome each of the five reasons why quality improvement efforts fail:

  1. Rate of change too rapid
  2. Amount of change too drastic
  3. Balance of consequences favors status quo
  4. Level of commitment too low
  5. Rate of progress seems too slow

- The single greatest impact you can have as a quality leader is to ensure that the balance of consequences your associates experience strongly favors quality improvement. (The steps to do this are explained in Chapter 10.)

- Managers may take a variety of actions to show leadership for quality in each of the three stages of implementation progress; these actions are listed in the following Leadership Actions Checklist.

- If you can only remember one thing about leadership, remember this:

"Setting an example is not the main means of influencing another, it is the only means."

—Albert Einstein

LEADERSHIP ACTIONS CHECKLIST

Stage 1

1. Ask yourself, "If my customers were standing here right now, what would *they* want me to do?"

2. Identify and document your *managerial* work processes (e.g., communication, delegation, recognition, staff development, coaching).

3. Identify the customers of your managerial processes.

4. Clarify the customer's requirements for each of your managerial processes.

5. Raise your group's quality awareness level.

6. Ask questions, make assignments, form teams, or do projects that cause all associates in your area to explicitly document customer requirements for the outputs of every job and process.

7. Explain (show how they apply) quality principles and concepts for each of your group's work processes.

8. Define the quality message you wish to send.

9. Use quality language in your everyday conversations.

10. Ask questions that demonstrate your understanding of, and commitment to, continuous improvement.

11. Get training on quality.

12. Read articles and books on quality.

13. Listen without making judgments.

14. Add quality topics to your weekly or monthly staff meetings.

15. Walk the talk.

16. Set the example.

17. Personally seek feedback from external and internal customers on a regular basis.

18. Ask yourself, "What have I done for the customer today?"

19. Ask yourself, "Are my words and actions sending the right message about quality right now?"

20. Above all, do whatever it takes to be a consistent champion for quality.

21. Define and establish the links between the elements of your QMS.

22. Demonstrate that you manage by fact, not by opinion. Describe and show the data used to reach your decisions.

23. Visibly review quality improvement plans and progress.

24. Spend time visibly improving processes yourself using your firm's continuous improvement process or tools.

25. Use the tools in Chapter 11 to help you integrate quality improvement actions into daily practice.

Stage 2

1. Explain to someone else how to use quality tools.

2. Explain to someone else how to manage by fact, not by opinion.

3. Explain that you expect everyone to manage by fact, not by opinion. Establish a ground rule that everyone may ask one another to describe and show the data used to reach decisions.

4. Spend time visibly explaining the processes that make up the QMS.

5. Increase the percent of time you spend on quality improvement actions.

6. Serve on a quality improvement team, especially if it's cross-functional.

7. Encourage employees to work actively on process improvement teams.

8. Strengthen the links between the elements of your QMS.

9. Look for success stories and make sure everyone hears the good news.

10. Review lessons learned from all improvement projects.

11. Encourage others to review lessons learned from all improvement projects.

12. Increase the frequency of the quality messages you send.

13. Use at least 50 percent of the total time you spend on quality improvement aligning the balance of consequences so that actions taken to improve quality are strongly reinforced.

14. Align performance measurement, recognition, and rewards so that they complement one another and encourage the actions necessary to meet customer expectations.

15. Communicate that you expect everyone to be a champion for quality.

16. Look for hidden or subtle examples of quality improvement (no matter how small) and make a big splash (i.e., stop the line when something right happens once in a while too!).

Stage 3

1. Hold improvement project forums to formally share and recognize project results.

2. Personally conduct those portions of new hire orientation that address the QMS and improvement process.

3. Verify that every employee in your area is able to answer the following questions:

    — Who is my customer?
    — What outputs does my job produce?
    — What value does my job add to my customer?
    — How does my customer measure the quality of what I provide them?
    — What standards must I meet in order to satisfy my customer?

4. Visibly take action to institutionalize the links between all seven elements of the QMS.

5. Maintain the frequency of your quality improvement communication.

6. Make heroes of everyone who is a quality champion.

7. Analyze how you are spending your time. Are you spending as much time as you feel you should on quality improvement? Is the time you spend focused on assuring and sustaining quality?

8. Examine your business planning process. Is quality improvement (customer satisfaction) fully integrated into your business plan?

9. Ask your associates to conduct an anonymous brainstorming session using the slip method or similar tool on the following topics: "What business as usual means around here"; "If I were president of this company for one day, here is what I would change"; "The biggest thing preventing me from satisfying my customer is . . . ." Reflect on the results of these sessions and ask yourself, "What does this data suggest I do next?"

CHAPTER FIVE

MANAGEMENT BY FACT

"No one is more definite about the solution than the one who doesn't understand the problem."

—Robert Half

EXECUTIVE OVERVIEW

- Management by Fact is data-driven decision-making.

- Two kinds of data are essential: customer satisfaction data (the voice of the customer) and process-performance data (the voice of the process).

- Total Customer Satisfaction, Quality Results and Process Quality Assurance processes provide the data used to make decisions (so that you can Manage by Fact).

- Management by Fact differs from business as usual.

WHAT IS MANAGEMENT BY FACT?

Management by Fact means using data and a systematic process to make decisions regarding how best to increase customer satisfaction.

WHAT DATA SHOULD YOU USE?

Two kinds of data are important: One is customer satisfaction data (the voice of the customer); the other is process-performance data, especially data related to process variation (the voice of the process). Without these two types of data, it is impossible to set customer-focused process-performance targets, establish predictive measures, measure process performance, or determine the root causes of quality problems.

WHERE DOES THE DATA COME FROM?

When you Manage by Fact, you use data generated (collected) from other parts of your quality management system (QMS) to make decisions. In other words, the output(s) of your data collection and measurement processes are inputs to your Management by Fact (decision-making) processes.

For example, as part of your Total Customer Satisfaction processes, you collect customer perceptions and expectations (external) data. One of your Process Quality Assurance processes establishes the measures you will use to assure that your work processes meet your customer's requirements. These process performance measures are used to collect (internal) data as part of your Quality Results processes.

Once you have the internal and external data generated from these parts of your management system, you use your Management by Fact (decision-making) processes to make decisions regarding the work processes in your area. This is the point at which the voice of the customer and the voice of the process come together to help you decide how best to satisfy your customers.

HOW DOES MANAGEMENT BY FACT DIFFER FROM BUSINESS AS USUAL?

Much of a manager's job involves making decisions. This is not new. What is new for many of us is the basis or methods used to make those decisions. Consider the following list of decision-making methods:

- Generate a list of alternatives; list the pros and cons of each alternative; assign relative weights to your pros and cons; pick the alternative with the greatest net score.

- Generate a list of alternatives; determine the cost and benefits of each alternative; pick the one with the greatest financial return.

- When faced with a decision, carefully consider the politics of the organization and pick the choice that is least sensitive politically.

- When faced with a decision, carefully consider the performance measures and compensation system of the organization and pick the choice that is most likely to reflect favorably one or both dimensions.

Chances are you've probably used one or more of these methods yourself, or know someone who has. The challenge you face when Managing by Fact is that you must overcome past habits or the tendency (if it exists) to use gut feeling, hunches, tradition, hearsay, past experience, superstition, or judgment as the sole basis for your decisions. Stated differently, if we are to Manage by Fact, we must broaden our repertoire of decision-making methods and tools. Table 5.1 shows how Management by Fact differs from business-as-usual decision-making:

TABLE 5.1

Business as Usual	Management by Fact
• We've got to do something right now (before things get even worse); quick fixes.	• Knowing when and when not to take action on a process.
• Treating symptoms; jumping to assumed solutions; acting on what appears to be correct.	• Reaching conclusions based on the result of root cause analysis.
• Focus on the ends not the means.	• Reaching conclusions based on understanding cause-and-effect relationships.
• Anecdotal customer information and hearsay.	• Reaching conclusions based on actual customer data provided by the Total Customer Satisfaction processes in the quality management system.
• Tunnel vision; zeroing in on a course of action that results in a low-yield improvement; focusing on what has worked in the past. "That's the way we have always done it."	• Developing a deep understanding of the sources of variation present in a process.
• *Believing* that a change is bound to make things better.	• Knowing *why* based on facts that a recommended course of action is preferred.

TABLE 5.1 (continued)

Business as Usual	Management by Fact
• We don't have time now to figure out why something isn't working; fire fighting.	• Using knowledge of cause-and-effect relationships to establish permanent fixes; fire prevention.
• Managing the pieces of the organization separately (i.e., the silo mentality).	• Viewing work in organizations as part of interdependent systems.

To Manage by Fact you will need to make use of the results (outputs) from additional decision-making methods and tools such as:

- Root cause analysis
- Scientific method
- Pareto analysis
- Cause-and-effect analysis
- Cost of quality analysis
- Return on quality analysis
- Continuous improvement process, *Kaizen,* or Plan, Do, Study, Act
- Benchmarking
- Disaggregation
- Process maps
- Checksheets
- Histograms
- Control charts
- Scatter diagrams
- Quality function deployment
- Design of experiments
- Stratification
- Experimentation

Additionally, whenever you first face a decision, ask yourself these questions:

- What data do I need to make this decision?

- Where will the data come from?

- Is the data valid? According to who?

- How will I use that data to select a course of action?

- What systematic process will I use to make the decision?

- If my customer was standing here next to me right now, what factors would he or she want me to consider? How do I know that for a fact?

Does Managing by Fact mean you have to be a know-it-all or a walking encyclopedia? Not at all! It simply means you have to develop the habit of asking the right questions and not moving on until you get the answers.

The following examples illustrate how managers at different organizations apply Management by Fact actions. As you read each example, ask yourself these two questions: What stage of implementation best characterizes this organization? and, Are the actions this manager takes appropriate?

EXAMPLE 1

Darren works for INTEK, a consulting firm that specializes in information technology. She manages a diverse group of technical hardware and software experts. INTEK has prospered as of late, primarily due to the widespread interest of businesses in more productively applying information technology to the management of cross-functional business operations. The partners in the firm believe there is an opportunity to gain even more new business if they can showcase how quality improvement processes and information management processes complement one another. Accordingly, each project manager has the responsibility to use quality tools where applicable as part of their client engagements. Darren and her group have completed INTEK's improvement process and tools training; they've also finished several in-house improvement projects over the last year.

During a session with a project team working on a client's employee expense reimbursement process, Darren finds that the team must help the client decide which type of error occurs most frequently on employee expense reports. The client has provided INTEK with copies of expense reports for the last six months. Darren poses the following questions to the team: "What decision do we need to make? What data do we need to make this decision? What process will we use to make this decision?"

EXAMPLE 2

An article in the August 8, 1994, issue of *Business Week* (primarily explaining return on quality) stated that Robert Allen, CEO of AT&T, receives a quarterly report from each of the company's 53 business units that spells out quality improvements and their subsequent financial impact. The article also states that based on AT&T's experience, when customers perceive improved quality, it shows up in better financial results three months later. "This is the most important thing that AT&T has ever done," Allen told a meeting of top managers the day before his June board presentation. The article goes on to say that to win approval from AT&T's top management these days, proponents of any new quality initiative must first demonstrate that the effort will yield at least a 30 percent drop in defects and a 10 percent return on investment.

EXAMPLE 3

In the same August 8, 1994, issue of *Business Week,* the author describes how, "Rethinking quality can force some companies to abandon cherished beliefs." He goes on to say that United Parcel Service Inc. had "Always assumed that on-time delivery was the paramount concern of its customers. Everything else came second." UPS recently began asking broader questions about how it could improve service and learned that customers wanted more interaction with drivers. "We've discovered that the highest-rated element we have is our drivers," says Lawrence E. Farrel, UPS's service-quality manager. "Now we're viewing drivers as more of an asset than a cost." As a result of this finding, the company is now allowing its 62,000 delivery drivers an additional 30 minutes a day to spend at their discretion to strengthen ties with customers and perhaps bring in new sales. Drivers are also being paid a small commission for any sales leads they generate. "The program has cost UPS $4.2 million in drivers' time so far this year, but has generated tens of millions of dollars in revenue," says Farrel.

Check your understanding of Management by Fact actions based on the following discussion.

EXAMPLE 1

Darren's consultants at INTEK are in Stage 1 of their QMS implementation. They've received training and completed some improvement projects. The questions Darren is posing to the team are appropriate actions for all three implementation stages whenever you face a decision.

EXAMPLE 2

You'll see this Stage 3 AT&T example in several places throughout the book because it does such a good job of illustrating the links that should exist between quality results data and customer satisfaction data, and how this data supports Management by Fact (decision-making) during one of your Strategic Quality Planning processes (identifying improvement opportunities). These actions are very appropriate in Stage 3 but would hinder most Stage 1 organizations because they typically do not have the data necessary to link quality improvements to customer satisfaction. What does this mean for Stage 1 organizations? Start by collecting customer satisfaction data and let the voice of the customer tell you what to improve!

EXAMPLE 3

UPS is an exceptionally well-run organization that appears to be in Stage 2. The decision to allow drivers to spend an additional 30 minutes with customers was based on the "new" customer satisfaction data it collected, even though this discovery contradicted a strongly held belief that on-time delivery was the paramount concern of its customers. To UPS's credit, it listened to the voice of the customer! This action is appropriate because it demonstrates a strengthening of the links between Total Customer Satisfaction and Management by Fact processes.

SUMMARY

- Management by Fact means using data and a systematic process to make decisions regarding how best to increase customer satisfaction.

- Your Management by Fact processes provide you with the answers to these questions:
  - Is increased customer satisfaction the driving force behind our decisions?
  - Are we using customer expectations, perceptions, and process performance data to make decisions regarding process performance, process improvement, and customer satisfaction?

— On what basis are performance goals, quotas, or standards being set?

— How do we know when—and when not—to change a process?

• Management by Fact differs from business as usual; it requires additional decision-making methods and tools.

MANAGEMENT BY FACT ACTIONS CHECKLIST

Stage 1

1. Decide *not* to take action to change a process unless you have data that tells you that change is necessary in order to improve the process (know when—and when not—to change a given process).

2. Read Section 3, "Managing in a Variable World," of Brian Joiner's book *Fourth Generation Management,* (McGraw-Hill, 1993).

3. Use outputs from the seven quality control tools to improve the effectiveness of your decisions.

4. Obtain customer expectations, satisfaction, process-performance (quality results) and root cause analysis data.

5. When considering a course of action, ask "What data tells us this proposed change will (eliminate the cause, improve customer satisfaction, reduce variation, decrease cycle time, etc.).

6. Listen to the voice of the process.

7. Ensure that your process improvement decisions are based on your understanding of the type of variation present in each of the work processes in your area.

8. Recognize that a trend without control limits is useless at best.

9. Visibly use quality tools, quality results data, customer satisfaction data, and root cause analysis data in decision-making.

10. Document the processes you currently use to make decisions.

11. Make a list of the types of data you typically use to make decisions.

12. When faced with the need to make a decision, ask yourself these questions:

- What data do I need to make this decision?
- Where will the data come from?
- Is the data valid? According to who?
- How will I use that data to select a course of action?
- What systematic process will I use to make the decision?
- If my customer were standing here next to me right now, what factors would he or she want me to consider? How do I know that for a fact? Is the (voice of the) customer represented in the decision-making process I plan to use for this decision?

13. Establish links between your Management by Fact, Total Customer Satisfaction, Strategic Quality Planning, Process Quality Assurance, Quality Results, and Human Resources Effectiveness processes.

14. Learn new decision-making processes.

Stage 2

1. Increase the use of data to help reach your decisions.

2. Identify the vital few pieces of data you require to manage your operation; develop reliable sources of this data so you may incorporate it into your decision-making processes.

3. Track (using your personal quality checksheet or other similar tool) and increase the percent of decisions you make based on data that confirms and demonstrates cause-and-effect relationships between process variables, process improvements, customer satisfaction, and financial impact.

4. Strengthen links between your Management by Fact, Total Customer Satisfaction, Strategic Quality Planning, Process Quality Assurance, Quality Results, and Human Resources Effectiveness processes.

5. Explain that you expect everyone to manage by fact, not by opinion. Establish a ground rule that everyone may ask one another to describe and show the data used to reach decisions.

6. Challenge your associates to track (using a personal quality checksheet or other similar tool) and increase the percent of decisions that the *group* makes based on data that confirms and demonstrates cause-and-effect relationships between process variables, process improvements, customer satisfaction, and financial impact.

Stage 3

1. Use the outputs from advanced quality tools such as Quality Function Deployment or Design of Experiments for decisions.

2. Institutionalize the links between your Management by Fact, Total Customer Satisfaction, Strategic Quality Planning, Process Quality Assurance, Quality Results, and Human Resources Effectiveness processes.

3. Calculate the percent of decisions made based primarily on confirmed customer satisfaction impact vs. cost, or productivity. Is this percentage indicative of a customer-focused QMS?

CHAPTER SIX

STRATEGIC QUALITY PLANNING

EXECUTIVE OVERVIEW

Strategic Quality Planning helps every manager:

- Deploy the voice of the customer throughout your part of the business.

- Plan to improve those processes in your area that most affect customer satisfaction.

The Sixty-Four Thousand Dollar Question Every Manager Should Be Able to Answer

How do you (or any other manager, for that matter) *know* what aspect(s) of your operation, department, or group on which to focus your attention? Past experience? Specific direction from higher-up? With all the things competing for your attention every day, wouldn't it be great if you had a foolproof way to know all the time what you should be working on and why? In an organization that is implementing a quality management system (QMS), here is how you would answer the above question(s).

What should I be working on (right now)? I should be working on those work processes that most impact the level of satisfaction that my internal or external customers experience. *Why* should I do that? In a well-designed QMS, focusing on customer satisfaction is like finding "true north" on a map. It's how we find our way and know where we want to go. Once we find it, we can't go wrong. *Until* we find it, we may feel like we're making progress, but we're really just going in circles. The part of the QMS that points the way on your quality improvement journey is Strategic Quality Planning.

WHAT IS STRATEGIC QUALITY PLANNING?

Strategic Quality Planning consists of the processes you use to develop specific plans to run your part of the business so that the work processes in your area that most impact customer satisfaction are identified, measured, improved, monitored, and managed.

As we stated in Chapter 3 (What Is A Quality Management System?), most managers mistakenly believe that Strategic Quality Planning is something that the senior executives do once a year for the entire enterprise. In reality, Strategic Quality Planning should become a vital part of every manager's job. Here's why. As part of a QMS, "Strategic" (as in Strategic Quality Planning) refers to the *impact* a given process has on customer satisfaction. The greater the impact on customer satisfaction, the *more strategic* the process is. Stated another way, the more *value* a given process adds to the customer, the more *strategic* that process is.

Thus, *your* strategic processes are those that add the most value to your customers, whether they are primarily inside or outside the organization.

The key issue for you as a manager is to identify which processes in your area are most strategic, and *why*? (Which have the greatest impact on customer satisfaction, and how do you know that for a fact?) How else do you know which processes or parts of processes to measure, monitor, improve, and manage in order to add the most value for your customer(s)?

INCORPORATE THE VOICE OF THE
CUSTOMER INTO YOUR PLANNING PROCESS

This chapter is not on how to plan, nor does it recommend a specific planning process. Rather, our intent is to suggest that you incorporate the voice of the customer into whatever planning process you use to establish your area's priorities to help assure the success of the operation.

To do this, you must first "translate" the voice of the customer data you obtain via your total customer satisfaction processes into relevant work processes and related process characteristics or variables terminology. Stated another way, you must define a clear link between the voice of the customer and the part of the operation that produces what the customer cares about most. For example, if on-time delivery is what your customers expect, then a relevant work process may be order entry, shipping, etc. Cycle time may be one of the related process characteristics or variables.

This is an area that a lot of companies simply don't do very well. One reason is that the data they collect from customers isn't actionable. It is too general to help identify what aspects of the organization's performance must change to increase customer satisfaction. Another reason some companies have difficulty is that they don't collect customer data—they assume they know what the customer wants as well as or better than the customer does. This happens often in high-technology companies or in companies that pioneer a new field and thus create the initial market.

Finally, some firms collect customer data but the data isn't made widely available throughout the business, or no one seems to know exactly who owns this information or how to obtain it!

How Do I Apply the Concepts of Strategic Quality Planning to My Daily Business?

Here are the steps to follow in order to incorporate Strategic Quality Planning into your current planning process.

1. *Determine the value that the processes in your area add to your customers.* Based on your knowledge of customer expectations data and your understanding of existing operations, can you describe the value your work processes create for or provide to your customers? This is more than a simple exercise in becoming customer-focused. Understanding the precise nature of the contribution that *your* work processes make to overall customer satisfaction is a necessary step if you are to measure, monitor, improve, or manage the right aspects of the work processes in your area. For example, are you in the business of processing orders or helping customers get merchandise in a timely fashion? If the answer is unclear, how do you know what to pay attention to in order to meet customer expectations?

2. *Identify which work processes in your area most impact customer satisfaction.* Does every work process in your area affect customer satisfaction equally? Or do some processes have a more direct impact than others? Ask yourself these two questions: If my customer was standing here right now, what part of the operation would they want me to focus on? Why is this important to them—that is, what value do they receive from this process?

   It is critical that you Manage by Fact when answering these questions. Your decision should be reached using customer data and a systematic process. Otherwise, the plans you create for these work processes will be based on untested assumptions, at best. If you singled out a work process to improve that had little or no impact on customer satisfaction and you were able to improve it dramatically, what good would it do?

The work processes that you identify during this step become your strategic processes for this customer or group of customers.

3. *Based on customer expectations data, define the desired process-performance targets or standards.* How well should the work process perform in order to meet customer expectations? For our on-time delivery example we must know what on-time delivery means to our customers so that we can establish internal process-performance standard(s) that will meet their expectations. If the customer defines on-time as within 24 hours, our processes might have to perform differently than if on-time means "within 30 minutes of your scheduled arrival time."

4. *Based on quality results data and customer perception data, determine if the strategic processes are currently satisfying customers.* Once you identify the specific processes that drive customer satisfaction and have established performance targets for each process, look at your quality results data to see how well each process is performing in relationship to the target. The gap between desired performance and actual performance becomes the basis for your process improvement objectives.

5. *Define your process improvement objectives.* What must you do to improve each strategic process? Once you know the level of improvement required to meet customer targets for your strategic processes, you can then plan process improvement projects (e.g., "reduce the cycle time of the shipping process by 50 percent.")

Your improvement objectives and process performance targets tell you what to pay attention to as a manager. They also give you a solid rationale for allocating resources among the processes in your area.

6. *Based on the results of Steps 1–5, develop a plan for the daily management, measuring, monitoring, and improvement of each of your "strategic" processes.* What actions will you take to improve each strategic process? What actions should others take? When will each action take place? Who will be responsible for each action? How will success be measured—in terms of progress and ultimate results? Your plan should direct the day-to-day activities necessary to ensure customer satisfaction is improving. Usually this involves continually examining and improving the performance of the work process, to reduce the amount of variation present and to increase the value the process adds to the customer. The processes that help you do this are your Process Quality Assurance processes. (See Chapter 7.)

The following examples illustrate Strategic Quality Planning actions of managers in three different organizations. As you read each example, try to identify which stage of implementation best characterizes the organization being described. Also, ask yourself, Are the actions this manager takes appropriate?

EXAMPLE 1

An article in the August 8, 1994, issue of *Business Week* (primarily explaining return on quality) stated that Robert Allen, CEO of AT&T, receives a quarterly report from each of the company's 53 business units that spells out quality improvements and their subsequent financial impact. The article also states that based on AT&T's experience, when customers perceive improved quality, it shows up in better financial results three months later. "This is the most important thing that AT&T has ever done," Allen told a meeting of top managers the day before his June board presentation. The article goes on to say that to win approval from AT&T's top management these days, proponents of any new quality initiative must first demonstrate that the effort will yield at least a 30 percent drop in defects and a 10 percent return on investment.

EXAMPLE 2

Chris manages a distribution center (warehouse) for a major retailer of consumer products. New to the company, Chris worked for a competitor whose name strikes fear into the hearts of many traditional retail stores because of its innovative distribution practices and state-of-the-art shelf replenishing practices. Brought on board to help accelerate the firm's quality improvement efforts in this part of the business, Chris has assembled the management team of the distribution center and asks the following questions: Who are our customers? What do they expect from us? What part of our operation most impacts their expectations? How do we know this to be the case for a fact? Chris looked around the room, saw the blank faces, and knew he had a challenge on his hands!

EXAMPLE 3

Phyllis is the manager of accounts receivable for a major telecommunications equipment manufacturer. Phyllis is a member of a cross-functional team that is responsible for the "health" of the firm's billing process. Customer satisfaction data indicates that customers feel the current bills they receive are too complex, often inaccurate, and arrive too late for them to take advantage of timely payment incentives. Process performance data for the invoice generation process show the cycle time to be 12 days from order entry to creation of the invoice. There are no performance targets in place for the invoice generation process other than the "average number of invoices generated per month of 5,000." Phyllis feels she should establish a timeliness target for invoices based on the customer expectations data.

Check your understanding of Strategic Quality Planning actions based on the following discussion.

EXAMPLE 1

This AT&T example describes an organization that is in Stage 3 of its QMS implementation. The action of creating a "minimum" return of 30 percent less defects and 10 percent on its investment is appropriate, since it has confirmed the relationship between *quality improvements perceived by the customer* and subsequent financial impact.

EXAMPLE 2

Chris's warehouse operation is early in Stage 1. He is attempting to identify which of the work processes in the warehouse are most strategic so he will be able to establish improvement priorities. This is an appropriate action for this stage.

EXAMPLE 3

This accounts receivable group is probably in Stage 1, though an argument could be made for Stage 2 since a cross-functional team has been established to monitor the "health" of the billing process. Establishing performance targets based on customer expectations data is appropriate for this stage.

SUMMARY

- Strategic Quality Planning processes tell you which work processes in your area to focus on and why.

- Your Strategic Quality Planning processes pull together data from your Total Customer Satisfaction and Quality Results processes; consequently they are only as effective as the links that exist between these QMS elements and the validity of the data provided by these links.

- Your strategic processes are the ones that add the most value to your customers.

- There are six steps to incorporate Strategic Quality Planning into your current planning process:

  1. Determine the value that the processes in your area add to your customers.
  2. Identify those work processes in your area that most impact customer satisfaction (*your* strategic processes).
  3. Based on customer expectations data, define the desired performance targets or standards for each strategic process.
  4. Based on quality results data and customer perception data, determine how well each strategic processes is currently satisfying customers.
  5. Define your improvement objectives for each strategic process.
  6. Based on the results of Steps 1–5, develop a plan for the daily management, measurement, monitoring, and improvement of each of your strategic processes.

STRATEGIC QUALITY PLANNING ACTIONS CHECKLIST

Stage 1

1. Integrate customer expectations, perceptions, and process performance data into your business planning process.

2. Identify the work processes in your area that most impact customer satisfaction; these are your strategic processes.

3. Seek customer and supplier input in the planning process.

4. Establish process-performance targets based on customer expectations data.

5. Identify process improvement objectives for each strategic work process.

6. Resist the temptation to improve quality for quality's sake; confirm the relationship between process performance and customer satisfaction before you improve a work process.

7. Determine and communicate what must be done daily, weekly, and periodically to operate processes so that they satisfy customers (i.e., daily management plans for each strategic process).

8. Review progress informally at regular intervals.

9. Align the existing balance of consequences for actions related to your strategic quality planning processes.

Stage 2

1. Determine precisely how every job in your area impacts customer satisfaction.

2. Determine or establish what the desired performance targets are for each job to meet customer expectations.

3. Use best-in-class benchmarks to set process-performance targets for those processes that are most important to your customers.

4. Set stretch goals to drive improvement (particularly in those processes most important to customers and that are likely to result in a sustainable competitive advantage).

5. Confirm whether and to what extent each of your associates is able to explain how their job impacts customer satisfaction, what the desired performance standards are to meet customer expectations, and how they measure process performance to ensure that the standard is being achieved.

6. Report on the status of process improvement objectives, process-performance targets, quality results data, and customer satisfaction data as part of all regular operations review meetings; prompt others to do the same.

7. Determine the economic impact resulting from process improvements by calculating the return on quality or similar measure.

Stage 3

1. Routinely discuss the results of improvement activities to determine whether the right processes were selected for improvement and the extent to which customer satisfaction improved.

2. Use return on quality or similar calculations to establish minimum success criteria (hurdle rates) for undertaking improvement projects.

3. Report on the status of process improvement objectives, process-performance targets, quality results data and customer satisfaction data as part of all regular operations review meetings; expect (require) others to do the same.

CHAPTER SEVEN

PROCESS QUALITY ASSURANCE

"The significant problems we face cannot be solved at the same level of thinking we were at when we created them."

—Albert Einstein

EXECUTIVE OVERVIEW

- Without an understanding of variation, you keep yourself needlessly in the dark.

- You must measure quality in order to manage quality.

WHAT IS PROCESS QUALITY ASSURANCE?

Author Stephen R. Covey tells the following story in his best-selling book *The 7 Habits of Highly Effective People:*

Suppose you were to come upon someone in the woods working feverishly to saw down a tree. "What are you doing?" you ask.

"Can't you see?" comes the impatient reply. "I'm sawing down this tree."

"You look exhausted!" you exclaim. "How long have you been at it?"

"Over five hours, and I'm beat! This is hard work."

"Well, why don't you take a break for a few minutes and sharpen the saw?" you inquire. "I'm sure it would go a lot faster."

"I don't have time to sharpen the saw," the man says emphatically. "I'm too busy sawing!"

Just like the person sawing down the tree in the woods, most managers say that they are so busy getting the work done, that they don't have time to improve *how* the work gets done. This is where your Process Quality Assurance processes can really help. They are the "sharpen the saw" part of the quality management system (QMS). These are the processes where the "technical" aspects of quality improvement take place—that is, those aspects that gradually improve the performance of a work process, such as continuous improvement, *Kaizen,* or Shewhart's PDSA cycle, as well as those that are intended to produce quantum leaps in process performance, such as benchmarking or reengineering. Process Quality Assurance processes also help you define, establish, and deploy the measures you use to assess and monitor how well your work processes are performing.

VARIATION AND MEASUREMENT

In our work with managers in general, and those managers who work outside of manufacturing in particular, we've discovered widespread misunderstanding of two critical concepts: Variation and Measurement. Yet, our experience is that these are two of the most important concepts managers must understand if they are to make progress at quality improvement, wish to Manage by Fact, or want to use the power of the scientific method to make better decisions and obtain better results in their part of the business.

First, we'll take a closer look at variation, then, later in this chapter, we'll present three key measurement questions that your Process Quality Assurance processes should address.

WITHOUT AN UNDERSTANDING OF VARIATION, YOU KEEP YOURSELF NEEDLESSLY IN THE DARK

Consider the following quote from Brian Joiner: in *Fourth Generation Management* (McGraw-Hill, 1994):

"Our ability to produce rapid, sustained improvement is tied directly to our ability to understand and interpret variation. Until we know how to react to variation, any actions we take are almost as likely to make things worse, or to have no effect at all, as they are to make things better" (p. 107).

Or this one from Dr. Lloyd S. Nelson:

"Failure to understand variation is a central problem of management" (p. 108).

What is variation, and why is it so important?
We will begin to answer the question in this chapter. For a more detailed explanation, please refer to Part Three (Managing in a Variable World) of Brian Joiner's excellent book.

WHAT IS VARIATION?

Variation is the natural differences in process outputs. Here are a few key points to remember about variation:

- There are **two** types of variation: variation that is inherent to every process, known as **common** causes of variation, and variation that is sporadic and usually comes from *outside* the process, known as **special** causes of variation.

- Common causes of variation are *always* present in a work process; they come from some combination of one or more of the following five categories: people, material, machines, methods, or measurement.

- Special causes of variation are *not* always present in a work process.

- Variation does not mean defects; defects occur when the amount of variation for a given output exceeds what the customer expects.

- Zero defects does not mean zero variation.

- There is no way to determine whether the variation present in a process output is due to common or special causes of variation simply by visually inspecting the output.

- The only way to determine the type of variation present (common or special) in a work process is to collect data on the outputs and chart and interpret the data using a suitable control chart.

WHY IS VARIATION SO IMPORTANT?

- *All* measures are subject to common and special causes of variation.

- The type of variation present (common or special) in a work process determines the type of improvement strategy that is most effective.

Let's look more closely at these two points.

All Measures are Subject to Both Causes of Variation

This may be one of the most important quality management principles you will ever encounter, because once you understand it and recognize it in your day-to-day work, you will not only dramatically improve your ability to manage by fact, but you'll also be able to identify cause-and-effect relationships more quickly and accurately.

Suppose you are a manager in a manufacturing plant that produces screws (see Figure 7.1). In order to sell the screws your plant makes, the screws have to be a certain length and possess threads of a particular width.

The screws are made by machines. Each machine produces hundreds of screws a day. Your plant is full of these machines. Each machine is operated by a skilled technician who inserts the raw material (such as stainless steel), adjusts the settings on the machine for the desired dimensions, and so on.

If you compare one screw to another, will they be *identical?* No. The screws will vary in some way, though it may not be visible to the naked eye. With the right measuring device, you can measure the finished screws as they are produced.

Will the measurement results be *identical?* No. Some values will be the same, some higher, and some lower. Why? Because all processes contain variation. Some examples of *common* causes of variation in the process that would affect the finished screws might be differences in the raw material used by each machine, ability of the machines to hold their settings consistently throughout the day, and wear and tear of the machines. See Figure 7.2.

FIGURE 7.1   Screw Manufacturing Process

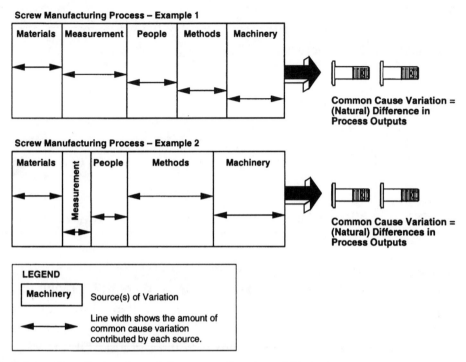

FIGURE 7.2   Two Examples of Common Causes of Variation

What if *special* causes were present in the process? Suppose a thunderstorm caused a power outage? Or the company starts using a new supplier of stainless steel whose material doesn't work as well in the screw machines? Perhaps one of the screw machines malfunctions. Could any of these affect the finished screws? Most certainly. See Figure 7.3.

Using Table 7.1, which shows the length of the first 76 screws produced by one machine at the plant, consider the following question carefully. Can

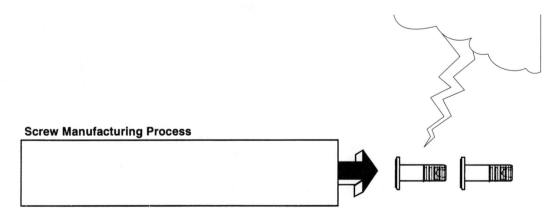

FIGURE 7.3   Special Cause of Variation (Power Outage)

TABLE 7.1

Screw Length in Centimeters

1.76	1.79	1.85	1.70	1.79	1.79	1.80	1.84	1.79	1.81
1.73	1.85	1.76	1.77	1.78	1.76	1.73	1.74	1.77	1.81
1.82	1.80	1.82	1.75	1.82	1.82	1.77	1.88	1.75	1.82
1.85	1.78	1.82	1.75	1.77	1.83	1.82	1.70	1.75	1.79
1.83	1.84	1.83	1.77	1.74	1.85	1.79	1.76	1.82	
1.79	1.80	1.79	1.80	1.76	1.73	1.76	1.73	1.83	
1.85	1.79	1.79	1.78	1.85	1.83	1.78	1.78	1.87	
1.80	1.81	1.80	1.80	1.84	1.81	1.79	1.84	1.79	

you tell whether the variation that occurred (differences in screw lengths) was caused by common causes, special causes, or both? (Recall that common cause variation is always present whereas special cause variation may or may not occur in the work process.) What if you chart the data? Perhaps that would help you better visualize what was happening in the process (see Figure 7.4).

The graph in Figure 7.4 helps you more easily see that screw length varies from a low of 1.70 centimeters to a high of 1.88 centimeters. But does this data help you tell whether the variation that occurred (differences in screw lengths) was caused by common causes, special causes, or both?

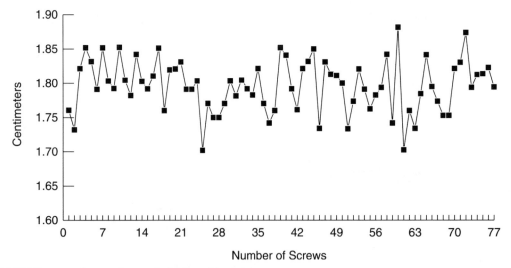

FIGURE 7.4   Screw Length in Centimeters

Process Quality Assurance   **83**

The correct answer is there's no way to tell from this data. We simply do not know from time series or trend data alone whether the process that produced these screws has only common cause variation, some special causes of variation, or a combination of both. Why not? Clearly there is some type of variation present or each screw would be exactly the same length.

What difference does it make? Plenty, if your goal is process improvement. Earlier we said that *all measures are subject to both types of variation.* You simply cannot tell which type of variation is present by looking at the measurement data points, even when they are plotted in time series order. We have no basis on which to improve the process, even though we can clearly see that from one screw to the next, the length definitely varies. Though we can make changes to the process, *we may not be improving the process.* Why? Because of the second principle of variation: *The type of variation present determines the appropriate improvement strategy.*

Consider the following possible courses of action for the process that is making these screws. Which would you choose?

1. Call a meeting of all machine technicians that produce screws on this machine. Show them the chart you prepared and explain that the goal is to reduce the variation in screw length by 20 percent. Offer a bonus to every technician that achieves the improvement goal.

2. Adjust the settings of this machine whenever a screw is produced that differs by more than .03 centimeters from the previous screw.

3. Check the settings on this machine. Reset the knob that controls screw length after each screw is produced, even though this slows down the total number of screws produced.

4. Collect data on this machine by technician, shift, and raw material. Look for patterns in the data.

5. Retrain the technician that operates this machine.

Before you decide, what if you were **certain** that the process that produced these screws was stable, predictable, and in control. In other words, only *common causes* of variation were present. Would that affect the choices you would make? How about if you knew that when a process is stable, there are certain types of strategies that are always recommended, and certain types that, if used, will make the process worse?

What if you were **equally certain** that the process was *not stable*—that is, one or more *special causes of variation were definitely present.* And you *knew* that when a process has special causes, there once again is a preferred strategy?

TABLE 7.2

If Type of Variation Is:	Then, the Appropriate Strategy[1] Is:	
Don't Know or You Aren't Sure	1. Collect process data. 2. Select appropriate control chart type. 3. Construct and interpret the control chart (chart will tell you whether variation present is due to common or special causes).	
Due to Special Causes	1. Get timely data so special causes are signaled quickly. 2. Put in place an immediate remedy to contain any damage. 3. Search for the cause; see what was *different* 4. Develop a long-term remedy.	
Due to Common Causes	1. Stratify	Group data by category; look for patterns in the data.
	2. Experiment	Make a small change on a pilot basis and study the results.
	3. Disaggregate	Break the work process down into subprocesses; select and use the appropriate strategy for each subprocess.

[1]The information in this table is based on the work done by Brian Joiner, from pp. 138, 139, 141, 147–153, and 259–269 of *Fourth Generation Management* (Brian Joiner, McGraw-Hill) © 1994 Joiner Associates Inc. All rights reserved. Reprinted by author with permission.

Table 7.2 summarizes the type of strategies you should take depending on whether common or special causes of variation are present. It is based on the work done by Brian Joiner, Ph.D., as presented in his book, *Fourth Generation Management.*

Now, read the five options again assuming that the screw manufacturing process was subject to only *common cause variation*. Which options are appropriate? Refer to Table 7.2 to help you decide.

Choice #1, #2 and #3 are examples of what Joiner calls tampering or overreacting to variation. Each is an attempt to adjust something in a system that is already stable (i.e., it is working as well as it is designed to work). As Joiner says, "In a common cause situation, there is no such thing as *the* cause. It's just a bunch of little things that add up one way one day, another way the next."

Choice #4 is an example of stratification—the only common cause strategy you have available to choose.

Choice #5 is a special cause strategy. It makes sense if you have already established that different results are occurring from technicians that are similar in all respects except the training they have received.

Here are two more situations that illustrate the power of knowing the type of variation present before you decide how to improve a process. These two situations have been reproduced in their entirety from pages 157 to 161 of *Fourth Generation Management* (Brian Joiner, McGraw-Hill) © 1994 Joiner Associates Inc. All rights reserved. Reprinted by the author with permission.

CASE #1: UNPAID INVOICES

Camille, the manager of accounts payable, walked through the department, pleased by the hum of activity. She entered her own office and got settled at her desk. A few minutes later Carlos entered, hesitating a bit at the door.

"What's up, Carlos?" she asked.

"Well, I just finished the figures for last month . . ." he said.

"And how do they look?"

"Uhhh, you're not going to like this . . ."

Camille steeled herself, then said, "Just give it to me straight. What's the news?"

Carlos checked the paper he was holding. "We're having a problem getting our suppliers paid on time. We had more than five percent of invoices unpaid last month."

"Over five percent! We haven't seen that in a long time. We used to hold invoices to help our own cash flow, but we're not doing that any more. What's going on out there? Let me look at those figures."

Look at the chart of Camille's figures on unpaid invoices. Here are the options she is considering. What would you recommend?

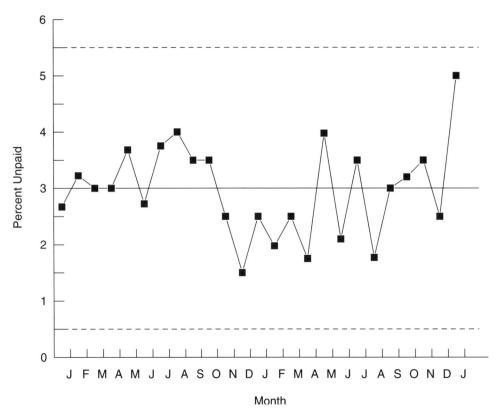

FIGURE 7.5   Data on Unpaid Invoices

1. Look at each invoice that was not paid and find out who worked on that invoice. Have the employee involved go through further training.

2. Figure out what was different last month compared to other months. Were there more invoices? Were new or different services being paid for the first time? Were there new employees in the department?

3. Dig up all the invoices that had to be reprocessed in the past few months and categorize the causes of the problem. Look for patterns.

4. For several weeks, have people working on each major step in the process keep data on how many and what kinds of errors occur in their steps.

5. Change the accounts payable software.

Hint: This control chart shows the invoice process to be stable. Refer to Table 7.2 and review options 1 to 5 in light of this knowledge.

The chart of Camille's data shows that this is a stable process, that there is no evidence of special causes. This means her best options are #3 and #4; they both represent common cause strategies.

Option 1 is tampering: Camille needs to look for patterns across all the data. Training would be appropriate only if she found systematic differences between people who were trained differently. Option 2 is a special-cause reaction: it assumes something was different or important about the last data point, which the control chart shows was not the case. Option 5 is a potential reaction but is premature at this point. Camille's available data do not show that the software is the cause of the problem.

Long-held beliefs are hard to overcome. Experience has shown that most people when presented with a situation like this are tempted to hedge their bets. They'll use one of the common cause approaches, but also ask the special cause question, What was different last month? this is particularly tempting when data points get near the control limits. All our instincts scream: "DO SOMETHING! Things will get worse if we don't act!" On some occasions, a data point near the limits may indicate the presence of a special cause, but *far more often* these points are due to common causes, and subsequent data points fall back well within the limits.

Now, let's look at the second situation:

CASE #2: INCREASED IMPURITIES

"Del, do you have a minute?"

"Sure, Aaron," said Del. "What can I do for you?"

Aaron handed Del a lab report.

"Eileen just finished measuring the impurities in the batch we're about to load into the trucks, and it doesn't look good. We've got an impurities measure of .23%. I know that we've got customers waiting for this shipment, but that's way above what we usually ship out."

"Let me think a minute," said Del. "That's more than one-and-a-half times what we had with the last batch. I thought we had been doing a good job of getting our impurities down."

Del paced the room several times then turned to Aaron. "Call the people waiting for this stuff and see if they can still use it. Otherwise, we may have to scrap the whole batch . . . I also want to see our data for the past few months so we can decide what else to do."

Look at Del's data in the chart. Here are the options he's considering. What would you recommend?

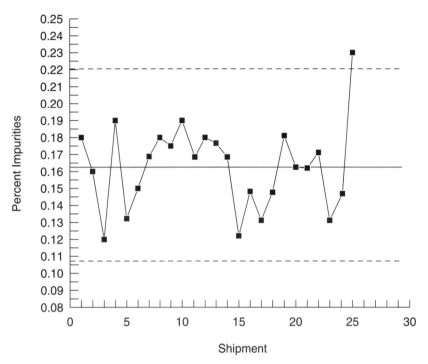

FIGURE 7.6   Data on Impurities

1.   Blend material from this batch with material from a different batch that has a lower impurity level. The combined mixture will yield an acceptable level of impurities. This will save the material from being scrapped.

2.   Look at the data by shifts, days, sampling machines, vessel types, and so on. See if there are patterns.

3.   Send out a memo to all staff members with information regarding the higher than normal impurity level. Note that there has been a bad batch and that operators should be more careful in the future. Post the control chart with the memo in the processing area so that all employees can see what is being shipped and what is expected of them.

4.   Make some educated guesses about possible causes, and change settings on the machines to see what happens. Have all pipes and vessels cleaned. Monitor the effect.

5.   Have employees investigate what was different about this batch relative to other batches. Was this batch processed differently? Were the measurements taken differently? Is there a new supplier?

Hint: This control chart shows that the process has at least one special cause present. Refer to Table 7.2 and review options 1 to 5 in light of this knowledge. Which options are appropriate when special causes are present?

The best option for Del is #5, which represents the only special cause approach here. The control chart shows that the last impurity measure was well beyond the control limit: something had changed the system. Option 1 might be necessary but could be dangerous because that action alone will not solve the problem and customers may suffer unintended side effects that you may or may not ever learn about. Options 2 and 4 represent common cause strategies (stratification and experimentation, respectively); they would help you find factors within the system but would not show you what was different in the last batch. Option 3 is perhaps the worst option: improvement does not come from exhortations or slogans. Until we know what has happened in the process, employees are likely to end up tampering in an attempt to make improvements.

> If you have looked carefully at the chart of Del's data, you may have also noticed that there was a special cause—a series of eight points above the centerline (points 7 to 14)—that appeared earlier and went undetected because no one was charting the data. If Del is lucky, he may still be able to track down the source of that special cause, and if he's luckier, it may be related to the latest event. But his chances now are much slimmer than if he had known when the special cause appeared and been able to initiate an immediate search for the cause. The trail grows cold quickly.
>
> It's useful to pause a moment and think about what our reactions might have been in these two cases had we not charted the data. Then think about the work environments throughout our organization. On how many processes do we have no data at all? How many times do people look at just the latest result in isolation? Just the latest two results? Appropriate use of figures requires charts like these.

Our thanks again to Brian Joiner for letting us use these two cases. Up to this point, we have focused on variation or, "the natural differences in process outputs." We started out by saying that all processes have some type of variation present, either common or special. This was important because knowing which type of variation is present (common or special cause) in the work processes you manage helps you make better process improvement decisions. As a reminder, the *only* way to know which type of variation is present is to collect process performance data and then chart and interpret the data using a control chart. Though we did not present any information on how to select, construct, or interpret control charts, we did show you two examples of completed control charts along with the relevant interpretation (special causes present or only common causes present) so that you could see how to use these tools. We also included a table to help you select the appropriate improvement strategy to use once you know the type of variation present.

Our intent in the previous section was to stimulate your interest in learning more about variation, control charts, and how they both can help you make more effective business decisions—that is, help you Manage by Fact. Hopefully, you'll take the time to seek out and read Joiner's book to learn more about these powerful concepts and tools. In the meantime, we now turn our attention to the second frequently misunderstood quality improvement concept, Measurement.

MEASUREMENT

How do you know how well a work process is performing? The simple answer is, measure it. Okay, then, perhaps we should ask, how do you know *what to measure* in the work processes you manage?

This is one of the key questions that your Process Quality Assurance processes should address. Two others are:

- What measures should we have in place to help predict customer satisfaction?

- Have we identified the critical factors within our work processes to measure and manage in order to produce (cause) high levels of customer satisfaction?

Note that one end result or output that you should look for from your Process Quality Assurance processes is that the right measures are in the right places in your work processes. The actual measurement data that you **collect** comes from your Quality Results processes (see Chapter 9, "Quality Results"). Stated another way, here, as part of your Process Quality Assurance Processes, you *establish* and *put the right measures in place.* Once they are in place, your Quality Results processes *make use* of the measures to collect and provide quality results data so that you can Manage by Fact, do strategic quality planning, assess customer satisfaction, etc.

What Should You Measure in a Work Process?

Believe it or not, our experience shows that this question is seldom asked or asked frequently enough. The answer we often hear is, "Whatever my boss asks me," or "Whatever I have to do get my bonus." Rarely do we hear the answer, "We should establish and tie our process performance measures directly to what our customers expect from this process." Yet, isn't this the single most important thing that you could hope for from your measurement data? Imagine being able to predict the level of customer satisfaction that your customers will experience by referring to a few select indicators or measures of performance while your process is operating! This is what leading firms are doing today. When their real-time process performance measurement system works well, the customer satisfaction data they collect *confirms* what they already know!

For most firms, however, the existing measures are financial or productivity oriented. Many support or administrative processes outside of manufacturing or operations may not even have measures in place. Try this simple test. Ask yourself or a fellow manager these two questions: What was the level of defects per unit and average cycle time for the———process last month? Based on what data?

At least one of these two measures always impacts customer satisfaction (i.e., customers want "it" right and on time, whatever "it" turns out to be).

In many firms, existing measures count or monitor things that have little value to end customers. For example, many managers run their departments with business plans or budgets. The manager receives a monthly report showing how much was spent against plan. Or the manager monitors productivity data. For example, a manager in a customer service center might be monitoring the number of calls received per hour. Is there anything wrong with these types of measures? No, of course not. However, if you run your business on these types of measures *alone,* you will have a very limited view of the business and little or no insight into what you should pay attention to if you want to maintain or increase the level of satisfaction that your customers experience.

Table 7.3 shows how business-as-usual measures compare with those developed as part of a customer-focused QMS.

What Measures Should We Have in Place to Help Predict Customer Satisfaction?

This is closely related to our first question, but in this case, the emphasis is on *defining the right measure* to assess and predict if a given expectation will be met. Our first question helps us figure out what to measure; this one prompts us to figure out *how* to measure that which our customers consider important.

This may sound simple but it can be quite challenging. For instance, suppose your customers think that the appearance of your product is important. What do you measure *while the process is operating* so that the finished

TABLE 7.3

Business As Usual	Total Quality
• Product measures and financial measures only	• Process and service measures with other business measures
• Meeting set targets	• Continuous improvement; goal continues to move up
• Verbal communication from customers	• Multiple sources of customer data
• Informal, or poorly defined, measurement system	• Organized, clear links to the other parts of the business
• Quantity measures, such as volume	• Process measures of effectiveness and efficiency
• Measures based on engineering specs	• Measures based on internal and external customer requirements
• Lagging historical measures	• Leading measures that predict future quality levels and customer satisfaction
• Measures focus on the individual (we assume the individual can control all the results)	• Measures focus on the process (we assume the process controls the results)

Source: Adapted from *Business Process Improvement* by H. James Harrington (McGraw-Hill, 1991)

product's appearance will meet or exceed your customers' expectations? One hint, it isn't cost per unit or total units produced per shift.

Two pieces of information are needed. One is knowledge of the specific aspects of appearance that your customer is concerned with, such as fit and finish, color, shininess, etc. This is voice of the customer data (customer expectations) that comes from your Total Customer Satisfaction processes (see Chapter 8). Without this data, you will be unable to Manage by Fact (see Chapter 5) or do Strategic Quality Planning (see Chapter 6). The second piece of information you need to define the right measures is knowledge of the (causal) factors in your work process that affect each aspect of appearance. This causal information is the output (result) of the processes you use to answer this question:

Have We Identified the Critical Factors Within Our Work Processes to Measure and Manage in Order to Produce (Cause) High Levels of Customer Satisfaction?

The processes that help you identify the relevant factors and establish the effect, if any, that each has on appearance or some other customer expectation are usually referred to as cause and effect analysis, root cause analysis, causal analysis, design of experiments, multivariate analysis, etc. They are all subsets of process improvement strategies, which, in turn, occur as part of your Process Quality Assurance processes.

Think back to the discussion of variation. Common cause variation will always be present, and thus will impact appearance in some way. Only by understanding the sources of variation and the effects each source has on a given aspect of appearance will you be able to monitor and manage the work process so that it produces products whose appearance consistently pleases your customers.

By using some combination of the three common cause process improvement strategies, that is, stratification, experimentation, and disaggregation, you will be able to identify the specific factors in the work process that cause the desired aspects of appearance. Then you create one or more measures for each factor so you can monitor that factor while the process is operating. Now you know what to pay attention to and why in this work process!

The following examples illustrate Process Quality Assurance actions of managers in three different organizations. As you read each example, try to identify which stage of implementation best characterizes the organization being described. Also, ask yourself, "Are the actions this manager takes appropriate?"

EXAMPLE 1

Phyllis is the branch manager for a large northeastern bank. This bank is well into its fifth year of quality improvement and intends to resubmit a Baldrige Award application. It narrowly missed the award last year, and has made dramatic improvements since that time in the few areas that were deficient. Phyllis wants to make sure that the various types and sources of data she uses to assure process quality and satisfied customers are well integrated. She used customer satisfaction data to set process performance targets, and developed measures for each work processes based upon those targets as well as the customer expectations themselves. She also identified the critical variables in each of her work processes that most impact customer satisfaction, and has measures in place where needed to monitor each of those variables while the processes are operating. Quality results (process performance) data is used to determine the effects of variation on each process; this data is used to identify when, and when not, to take action on each work process. She has also incorporated the data obtained from benchmarking projects to help establish improvement objectives for selected work processes. Finally, she draws upon all of this data to provide feedback to her associates.

EXAMPLE 2

Carl manages the employee accounts payable department for a large midwestern company. Recently, the firm has adopted a company-wide continuous improvement process. Carl has attended the company's training on the improvement process and associated "basic" tools. Carl is curious about how the improvement process might apply to his work area, so he decides to see if he can apply what he learned by examining the employee travel expense reimbursement process. He has received several complaints lately regarding the time it takes for employees to get reimbursed for trips they take on behalf of the company. As Carl begins his investigation, he finds that very little data is available on the time it takes to process expense reports. The only things currently being measured are the number of expense reports processed per month and the total dollar amount of the expenses reimbursed monthly. By talking with his associates, he finds that some expense reports take longer than others to process for various reasons. It seems that everyone has a different idea of how long it takes to process a typical expense report. Intrigued, Carl decides to begin collecting data on the time it takes to process expense reports so he can better understand what is actually occurring in the travel expense reimbursement process.

EXAMPLE 3

Oscar is a member of a cross-functional team working to improve his company's order entry process. This is one of several improvement projects Oscar has been involved in since his firm began its "quality journey" three years ago. The team is early in the project when they review data that shows that some offices in the company enter orders more accurately and quickly than others. One of the team members suggests that the full team single out these offices for close examination in order to find out what they do differently from the other offices. Oscar says, "You may be right, but it seems that we should first determine whether we are seeing the effects of common or special cause variation so we can select the appropriate improvement strategy."

Check your understanding of Process Quality Assurance actions based on the following discussion:

EXAMPLE 1

Phyllis's branch is in Stage 3 of its QMS implementation. It has all the elements in place (based on narrowly missing the Baldrige Award last year) and has had an additional year to strengthen its system. The actions Phyllis has taken to integrate the types and sources of data is very appropriate for this stage. These actions are necessary to institutionalize the links between your Total Customer Satisfaction, Strategic Quality Planning, Process Quality Assurance, Quality Results, Management by Fact, and Human Resources Effectiveness processes.

EXAMPLE 2

Carl's employee accounts payable department is in Stage 1. He has recently completed the company's improvement process and tools training. There is no formal improvement project yet, but Carl's curiosity is well placed. His decision to begin collecting data is sound at this stage.

EXAMPLE 3

Oscar's firm is probably in Stage 2. They've been doing improvement projects and have established at least one cross-functional team to improve the order entry process. Oscar's suggestion to determine the type of variation present in the order entry process and then select the appropriate improvement strategy is right on target.

SUMMARY

- Process Quality Assurance processes are where the "technical" aspects of quality improvement take place, that is, those that gradually improve the performance of a work process, such as continuous improvement, *Kaizen,* or Shewhart's PDSA cycle, as well as those that are intended to produce quantum leaps in process performance, such as benchmarking or reengineering.

- Process Quality Assurance processes also help you define, establish, and deploy the measures you use to assess and monitor how well your work processes are performing.

- The measures that you develop or have in place produce the data that your Quality Results processes collect; thus, there is a critical link that must be established between these two elements of the QMS.

- At least some of the measures should be based on the process performance targets you established during your Strategic Quality Planning processes; thus, there is a critical link that must be established between these two elements of the QMS.

- Some of the measures that you develop or have in place should help you monitor critical process variables that impact customer satisfaction, and show you whether special or common cause variation is present so you can better Manage by Fact.

- Process measures should be used to provide feedback as part of your Human Resources Effectiveness processes; thus, there is a critical link that must be established between these two elements of the QMS.

- Customer perceptions data is also used as part of your Process Quality Assurance processes to confirm whether improvement projects (process changes) achieved the desired result; thus, there is a critical link that must be established with your Total Customer Satisfaction processes.

- "Our ability to produce rapid, sustained improvement is tied directly to our ability to understand and interpret variation. Until we know how to react to variation, any actions we take are almost as likely to make things worse, or to have no effect at all, as they are to make things better."[1]

---

[1] Brian Joiner, *Fourth Generation Management*, McGraw-Hill, 1994, p. 107.

PROCESS QUALITY ASSURANCE
ACTIONS CHECKLIST

Stage One

1. Reread Einstein's quote at the beginning of this chapter.

2. Obtain the data you need from your total customer satisfaction processes (customer expectations and feedback data). If it doesn't exist, make getting it your top priority.

3. Read Section 3, "Managing in a Variable World," of Brian Joiner's book *Fourth Generation Management.* Tell your friends to do the same.

4. Use a systematic continuous improvement process whenever you recommend a change to a work process. Attend your company's training on this process.

5. Learn the seven quality control tools; read Kaoru Ishikawa's *Guide to Quality Control,* or whatever your firm's recommended equivalent may be. Attend your company's training on these tools.

6. Learn these processes for root cause analysis: Ask "why?" five times, (each answer gets closer to the root cause), Scatter Diagrams (help determine possible relationship between two factors), and Cause and Effect Diagrams (lists possible causes for each of the five sources of variation present in a work process).

7. Learn to recognize the sources of variation in all the work processes you manage.

8. Learn to recognize the four types of tampering that are prevalent in the work processes you manage or perform.

9. Help others learn to recognize the four types of tampering that are prevalent in the work processes you manage or perform.

10. Resist the temptation to tamper; better yet, resist the temptation to cause those who work for/with you to tamper.

11. The next time a proposed change to a process is discussed, ask, "Are we tampering, or will this change be an improvement? How do we know that for a fact?"

12. Analyze existing process performance measures:

    — Does the measure monitor a process characteristic that is important to the customer?

— Is there a measure for each customer's expectations?

— Is the measurement taken at a point in the process where it will prevent or detect defects?

— Do you have a measure for the quality of the inputs to this process?

— Is the input from the suppliers to this process usable as is?

— Do people know what to do, if anything, as a result of this measure?

13. Establish the links between your Total Customer Satisfaction, Strategic Quality Planning, Process Quality Assurance, Quality Results, Management by Fact, and Human Resources Effectiveness processes.

14. Create measures where needed to monitor and manage the variables that cause customer satisfaction.

15. Measure all work processes whether they produce products or services.

16. Eliminate measures that do not help you manage your work processes.

17. Align the existing balance of consequences for actions related to your Process Quality Assurance processes (particularly for actions related to quality improvement and measurement). See Chapter 10.

18. Evaluate the links between your Total Customer Satisfaction, Strategic Quality Planning, Process Quality Assurance, Management by Fact, Quality Results, and Human Resources Effectiveness processes.

19. Coach your associates so that they help create measures where needed to monitor and manage the variables that cause customer satisfaction.

20. Never use measures as a "club." Do not attempt to assign blame based on the data provided by measures.

Stage Two

1. Ask yourself this question often: Am I seeing the effects of common or special cause variation?

2. Collect process performance data and use it to determine whether process variation is due to special or common causes in each of your strategic (high impact on customer satisfaction) processes.

3. Learn how to use the right kinds of control charts for the work processes in your area.

4. Apply the appropriate process improvement strategies based on the presence of special or common cause variation.

5. Teach employees how to understand variation as it applies to their process measures and how to respond appropriately.

6. Build in measures for processes so that the associates who work in them receive real-time feedback directly from the process.

7. Strengthen the links between your Total Customer Satisfaction, Strategic Quality Planning, Process Quality Assurance, Management by Fact, Quality Results, and Human Resources Effectiveness processes.

8. Conduct or personally help someone else perform a root cause analysis on a key process that does not meet customer requirements.

9. Learn more complex processes for root cause analysis such as Design of Experiments (looks at interactions among several process factors).

Stage Three

1. Determine how current process performance compares to competition or to best-in-class performance, identify the relevant *metrics, practices, and* enablers of world class process performance for the strategic processes in your area (i.e., participate in a benchmarking project).

2. Do not change a process unless you have data that shows you the type of variation present in the process and the impact the change would have on customer satisfaction and profits.

3. Institutionalize the links between your Total Customer Satisfaction, Strategic Quality Planning, Process Quality Assurance, Quality Results, Management by Fact, and Human Resources Effectiveness processes.

4. Design or redesign your processes so the need to measure is reduced or eliminated (the reliability and robustness of the process is so consistent that most measures are no longer necessary).

CHAPTER EIGHT

TOTAL CUSTOMER SATISFACTION

"Companies succeed by providing superior customer value. And value is simply quality, however the customer defines it, offered at the right price. Superior customer value is the best leading indicator of market share and competitiveness. Market share and competitiveness, in turn, drive the achievement of long-term financial goals, such as profitability, growth, and shareholder value."
—Bradley T. Gale, *Managing Customer Value* (Free Press, 1994)

EXECUTIVE OVERVIEW

- The primary outputs of Total Customer Satisfaction processes are customer expectations data and customer perception data.

- Total customer satisfaction processes help:

  — Define internal and external customers for every work process.
  — Determine the expectations of those customers.
  — Determine whether and to what extent customers are satisfied.

WHAT IS TOTAL CUSTOMER SATISFACTION?

Customer Satisfaction is the degree to which customer perceptions exceed their expectations. Total Customer Satisfaction processes are the data collection processes used to obtain customer expectations and perceptions data.

When systems theorists analyze systems, they look for the aim or purpose of the system. The aim of a business system is what the organization

101

wants and needs to do to stay in business. Tom Peters argues that most companies are organized around and for the convenience of their functional hierarchies. For example, Peters mentions that hospitals are built around doctors, manufacturing firms are organized around production efficiency, banks around efficient backroom practices. The question Peters asks is "How many [organizations] build the entire logic of the firm around the customer?"[1]

In a quality management system (QMS), the aim of the system is customer satisfaction—in other words, the firm is built around processes designed to provide value to the customer. Total Customer Satisfaction is about applying Big Q thinking (quality that affects the entire company) to the business, not just little q (product quality).[2] Without Big Q, the organization will not stay in business. Improving Big Q means increasing the level of satisfac-tion that customers (both internal and external) experience throughout the business.

WHY IS CUSTOMER SATISFACTION IMPORTANT?

The voice of the customer focuses and drives the remaining elements of your QMS. Customer expectations data helps set performance targets as part of Strategic Quality Planning; this data is also used to create measures as part of Process Quality Assurance. Customer perceptions data tells you if you hit the target and allows you to Manage by Fact. There is no point in doing all the other things a QMS requires unless you have a firm foundation of customer knowledge. Without a deep understanding of customer expectations and perceptions, you are just managing from assumptions—in other words, it's business as usual.

When managers fail to explicitly acknowledge or incorporate the voice of the customer into their work processes, they run the risk of focusing their attention on a process that has little value to customers. At best, this means wasting resources that could be channeled into other activities; at worst, in these lean times, it could spell the demise of that work group.

Many companies are unhappy with their quality efforts because they are not getting the results they expect. Quality is measurably better but business isn't. Why? Because they are working on improving processes that are *of low value to the customer*. In other words, these firms underemphasize customer data and overemphasize quality results (process performance) data.

---

[1] *Customerizing*, Occasional Papers, Tom Peters, © 1991 TPG Communications.

[2] Big Q commonly refers to meeting or exceeding customer expectations. Little q refers to the relative excellence of a product or service in areas such as reliability or durability.

In a recent article in *Quality Progress,* John Goodman, the president of TARP, a customer research center in Washington, D.C., remarked:

> The bad name that TQM is receiving is a matter of execution, not intent. Well-intentioned executives think their TQM efforts are customer-driven, but they are not. Instead of being customer driven, many companies' priorities are based on management's perceptions of key customer problems (perceptions that are often wrong) or on poor interpretation of data on customer's problems. As a result, TQM starts with a rational objective and arrives at an irrational conclusion.[3]

Still not convinced? Then look at your in-basket. How much of your company mail is externally focused (customer-specific issues, customer measures or data) vs. internally focused (committee minutes, announcements, interoffice mail)? In essence, you are what you read. Tom Peters states, "If the information that flows by you and through you—memos, meetings, minutes—is largely internal in content, then I believe, so are you."[4]

WHY ISN'T CUSTOMER DATA MORE READILY AVAILABLE?

Most managers and their companies do not do a good job of collecting and using customer data. The reasons include:

- They already think they know enough about their customers ("I've been in this business for 20 years and I know what they want," or, "Our customers don't know what they want.").

- They believe it's too hard or too labor intensive ("We don't have time to talk to customers," or, "We can't afford an expensive customer survey.").

- The data they do collect is poorly dispersed or unused. For example, sales or service functions often have customer data but it's rarely shared with other parts of the organization.

- They collect data that is too general. Firms often invest in extensive surveys that produce findings such as "our customers want better ser-

---

[3]"The Key Problem With TQM," John Goodman, Gary Bargatze, and Cynthia Grim, *Quality Progress,* January 1994, p. 45.

[4]*Quality,* Occasional Papers, Tom Peters © 1986, TPG Communications.

vice." Based on this data, what would you work on first to provide better service?

HOW TO HEAR THE CUSTOMER'S VOICE

To sharpen your customer focus, you should answer these four questions for each of your work processes, beginning with those that are most strategic (see Chapter 6, "Strategic Quality Planning"):

- Who are the customers for this process?

- What do these customers expect from this process?

- How well are we meeting those expectations? Based on what data?

- What action is necessary for those expectations that we are not currently meeting?

Let's look at these in more detail.

Who are the Customers for This Process?

Just as every process has a product or service as its output, *every* process also has a customer.[5] In fact, most processes have many customers, from intermediaries to the ultimate end-user. Often, managers do not have a clear picture of their customers. This is especially true of managers who run internal support departments. A *department* is not a customer; the individuals within that department are. An often overlooked group of customers are *your employees.* These people receive the outputs from your *managerial* processes.

The more specific you are about your customers, the better off you'll be. For example, for years an electrical equipment manufacturer grouped its customers in four main categories: industrial, construction, utilities, and original equipment manufacturers. After conducting some further research, the company found these categories were far too broad and that customers' needs were much more segmented than originally thought.

---

[5]We will define a *customer* as "anyone who receives, purchases, or uses a product or service from a supplier." This can include outside customers or even the next operation on a production line. The important thing to remember is that we are all customers and suppliers at various times, linked together in a chain that leads to the external customer, or end-user.

What Do these Customers Expect From This Process?

Expectations are stated or unstated customer requirements. Clarifying expectations helps you establish performance targets that your company must meet in order to secure and keep the customer's business. Expectations data also helps you develop process performance measures. (See discussions in Chapter 6, "Strategic Quality Planning," and Chapter 7, "Process Quality Assurance," respectively.)

Give your customers a chance to tell you what they want. Let them tell you where you are winning or losing the fight. Asking good questions and listening in a consistent manner are essential prerequisites of a customer-focused manager.

Here are some questions to help surface expectations: What exactly do you want from the product or service? What's most important to you? Are you getting what you need? How do you use our product or service? What should be different? How can I make it easier for you to use our products or services?

Asking these types of questions can be as simple as a phone call or a face-to-face visit, or as elaborate as focus groups and surveys.[6]

How Well are we Meeting Those Expectations? Based on What Data?

The previous question provides you with expectations data. To fully understand the voice of the customer, you must also collect *perceptions* data. Perceptions are subjective assessments of your products, services, and company. Most perceptions relate to the value your customers feel they receive from your product or service. These perceptions are based on your customers' direct or indirect experiences, and even materials or literature. Perceptions provide feedback on how well you are doing in the customers' eyes.

The important thing to remember about the voice of the customer and perceptions data in particular is no matter how unpleasant these "facts" are, you must treat them as real. As Tom Peters says:

> For better or worse, your outfit is not real. It is no more, and no less, than the sum total of the . . . images created by your customers,

---

[6]For more detailed information on collecting customer expectations, see any of the articles by Valerie Zeithaml, A. Parasuraman, and Len Berry on their SERVQUAL surveys.

employees, vendors, distributors and communities as they experience you on a day-to-day basis.[7]

When asking about perceptions, you can query: "Did you feel the product/service was worth what you paid for it? Would you buy it again?" (Or, for internal customers, "If you had a choice, would you buy elsewhere?")

What Action is Necessary for Those Expectations That we are Not Currently Meeting?

Customer expectations and perceptions data are important inputs to the Strategic Quality Planning process. If there is a gap between what customers expect and what they perceive you provide, you would establish an improvement objective as part of your Strategic Quality Planning processes.

The improvement objective would trigger Process Quality Assurance processes, and you would use a combination of Quality Results data and customer perceptions data to confirm whether the improvement objective was met so you Manage By Fact.

To recap, customer expectations and perceptions data help you focus your managerial attention. They alert you to what is important, and confirm how well you are doing at managing these important things.

One final dimension of Total Customer Satisfaction: there is a strong relationship between internal customer satisfaction (the employees inside the business) and external customer satisfaction. Several years ago, a book titled *The Customer Comes Second* was published. The title shocked a lot of quality professionals, but the message of the book was sound: if you don't have satisfied employees, you will never have satisfied customers.

We began this chapter by stating that the aim of a QMS is customer satisfaction. Satisfied customers produce profit for the business, not vice versa. To learn more about the relationship between customer perceptions of quality and profits and the importance of collecting customer data, we encourage you to read *Managing Customer Value* by Bradley T. Gale.

The following examples illustrate Total Customer Satisfaction actions of managers in three different organizations. As you read each example, try to identify which stage of implementation best characterizes the organization being described. Also, ask yourself, "Are the actions this manager takes appropriate?"

---

[7] *Customerizing,* Occasional Papers, Tom Peters © 1991, TPG Communications.

EXAMPLE 1

Jerry manages a plastic molding operation. He recently chartered an improvement team to reduce defects associated with the molding process. This will be the first improvement project for these team members. Jerry wants to make sure that the team is working on issues that are important to customers, so he arranges for the team members to visit a big user of the finished product. The goal of the visit is to define and clarify the customers' expectations and get some first-hand information on how the product is being used.

EXAMPLE 2

An article in the August 8, 1994, issue of *Business Week* (primarily explaining return on quality) describes how United Parcel Service Inc., "had always assumed that on-time delivery was the paramount concern of its customers. Everything else came second." The article continues to explain that "the problem was, UPS wasn't asking its customers the right questions. Its surveys barraged clients with queries about whether they were pleased with UPS's delivery time and whether they thought the company could be speedier.

When UPS recently began asking broader questions about how it could improve service, it discovered that clients weren't as obsessed with on-time delivery as previously thought. The biggest surprise to UPS's management: Customers wanted more interaction with drivers, the only face-to-face contact any of them had with the company. If drivers were less harried and more willing to chat, customers could get some practical advice on shipping. "We've discovered that the highest-rated element we have is our drivers," says Lawrence E. Farrel, UPS's service quality manager. "Now, we're viewing drivers as more of an asset than a cost."[8]

EXAMPLE 3

According to the same article in the August 8, 1994, issue of *Business Week,* the author writes that "extensive surveying, perhaps even inviting customers into design and production processes, helps companies identify the key factors that affect customers' buying decisions." The author tells the story of "one company that is looking closely at

---

[8]*Business Week,* August 8, 1994, p. 58.

how and why customers choose to buy."[9] Promus Co., the Memphis-based hotel and gaming company that owns Hampton Inns and Embassy Suites, offered guaranteed refunds to any customer dissatisfied with their stays for any reason at a Hampton Inn. The program not only brought in an additional $11 million in revenue, but employee job satisfaction climbed steadily (*everyone* was empowered to grant the refunds). It also helped Promus identify and resolve guests' chief annoyances at its Embassy Suites Inc. chain, one of which was a lack of irons and ironing boards. "We have literally no problems now from an area that was one of the largest complaint generators," says Mark C. Wells, Promus's senior vice-president for marketing.[10]

Check your understanding of Total Customer Satisfaction actions based on the following discussion.

EXAMPLE 1

Jerry's molding process is in Stage 1 of its QMS implementation. The action to meet directly with customers and learn their expectations is very appropriate for this stage.

EXAMPLE 2

UPS is in Stage 2. It has already been surveying customers and working to improve quality, but it is strengthening its data collection processes by asking better questions. It is not unusual for companies to discover that what they thought customers wanted is different from what real live customers expect. The key is to be willing to ask the right questions and then act on what you hear, which UPS did.

EXAMPLE 3

Promus sounds like a Stage 3 firm, or at least one that's in the latter part of Stage 2. It is trying to get closer to its customers, both internal and external. It is using the customer data it collects to identify and resolve problems, like the one involving ironing boards, and is tracking its progress at eliminating the source of the complaints.

---

[9] *Business Week,* p. 56.

[10] *Business Week,* p. 57.

SUMMARY

- Total Customer Satisfaction processes are data collection processes.

- Two types of data should be collected: expectations data and perceptions data.

- Both types of data are essential to help focus your managerial attention: *Expectations* data tells you the minimum level of performance that customers expect from your work processes; perceptions data tells you how the results from your work processes compare with customer expectations.

- Collectively, these two types of customer data should feed into your Strategic Quality Planning, Process Quality Assurance, Human Resources Effectiveness, and Management by Fact processes.

TOTAL CUSTOMER SATISFACTION ACTIONS CHECKLIST

Stage 1

1. Identify your customers by name.

2. Obtain and clarify your customer's expectations.

3. Obtain and clarify your customer's perceptions.

4. Personally sponsor and encourage customer visits for your associates.

5. Help your associates see how their work results affect customers.

6. Work with every associate in your area so that they are able to answer the following questions:

   — Who is my customer?
   — What outputs does my job produce?
   — What value does my job add to my customer?
   — How do my customers measure the quality of what I provide them?
   — What standards must I meet in order to satisfy my customer?

7. Verify that the process performance targets and measures you use directly contribute to meeting customer expectations.

8. Personally visit another department, branch, or other company that is known for high levels of customer satisfaction. Identify and document what you learn.

9. Identify the customer listening posts (sources of customer data) in your company or department.

10. Establish and put in place a method to collect customer expectations and perceptions data on a regular basis.

11. Conduct your own customer satisfaction audit. For example, some hospitals insist doctors spend time as a "patient," seeing the hospital through a patient's eyes. In one hospital, the admitting department employees pretend to be incoming patients and are "checked" into the hospital. In a more formal way, some organizations employ mystery shoppers to use the service and provide feedback.

12. Align the balance of consequences so that they reinforce the collection of customer data.

13. Begin collecting customer data.

14. Identify the customer data currently available.

15. Identify the methods currently used to collect customer data.

16. Identify the methods currently used to compile, organize, and communicate customer data.

17. Assess what happens to the customer data you currently collect.

    — How is it used?
    — Who uses it?

18. Assess the timeliness, availability, or usefulness of the customer data you currently collect.

19. Evaluate the links between your Total Customer Satisfaction, Strategic Quality Planning, Process Quality Assurance, Human Resources Effectiveness, Management by Fact, and Quality Results processes.

20. Establish links between your Total Customer Satisfaction, Strategic Quality Planning, Process Quality Assurance, Human Resources Effectiveness, Management by Fact, and Quality Results processes.

Stage 2

1. Integrate customer information into company training programs, especially new hire orientation, management development, and quality improvement training.

2. Share and use customer data regularly in meetings; demonstrate that you are using customer data to Manage by Fact.

3. Improve the timeliness, availability, or usefulness of the customer data you currently collect.

4. Increase the use of customer data you collect.

5. Improve the methods used to collect customer data.

6. Improve the methods used to compile, organize, and communicate customer data.

7. Strengthen the links between your Total Customer Satisfaction, Strategic Quality Planning, Process Quality Assurance, Human Resources Effectiveness, Management by Fact, and Quality Results processes.

8. Use customer data to develop a service strategy as part of your Strategic Quality Planning processes. A service strategy describes how you want your customers to perceive you, and what actions you will take to improve perceptions.

9. Design or modify your work processes and jobs so that they directly provide customer expectations and perceptions data to associates while the process is operating (i.e., create a built-in feedback capability).

10. Coach employees in collecting and analyzing customer data.

Stage 3

1. Use advanced quality tools such as Quality Function Deployment or Hoshin Planning to integrate the voice of the customer into core processes.

2. Ensure widespread use of the customer data you collect.

3. Automate the methods used to collect customer data.

4. Automate the methods used to compile, organize, and communicate customer data.

5. If you haven't already done so, integrate your Total Customer Satisfaction processes and data into your company's overall information system capability.

6. Institutionalize the links between your Total Customer Satisfaction, Strategic Quality Planning, Process Quality Assurance, Human Resources Effectiveness, Management by Fact, and Quality Results processes.

CHAPTER NINE

QUALITY RESULTS

When you can measure what you are speaking about and express it in numbers, you know something about it, and when you cannot measure it, when you cannot express it in numbers, your knowledge is of a meager and unsatisfactory kind. It may be the beginning of knowledge, but you have scarcely in your thoughts advanced to the stage of science.

—Lord Kelvin

EXECUTIVE OVERVIEW

- Quality Results processes *collect* work process performance data.

- The primary use of this data is feedback.

- The data collected directly supports these other quality management system (QMS) elements:
  - Strategic Quality Planning
  - Process Quality Assurance
  - Management by Fact
  - Human Resources Effectiveness

WHAT ARE QUALITY RESULTS?

Quality "results" means performance data. Quality Results processes are the data collection processes that supply you and other parts of the organization with feedback on the performance of the work processes you manage. Collectively, your Quality Results processes should help you answer these key questions:

- What aspects of (work) process performance are we currently measuring? (Does the data you collect help you monitor the key variables and sources of variation within the process? Or does it primarily focus on the outputs of the process?)

- Do our (work) process measures help us manage the process, or are we simply tallying low-value data? (Does the data help you make better, more timely decisions? Does it help prevent defects, or simply count them so you can sort the good from the bad? Overall, is the value of the data you collect greater than the effort involved in collecting it?)

- How well is the (work) process performing? (Does the data inform or overwhelm you? Can you quickly identify how well the process is performing? Whether it shows common or special causes of variation?)

- What happens to the measurement data we collect? (Is it readily available to everyone who works in the process? Shared with other parts of the organization? How is the data used? To provide feedback? Make process improvement decisions?)

WHY IS QUALITY RESULTS DATA IMPORTANT?

Quality Results data helps you:

- Monitor and manage process performance.

- Obtain feedback on the *results* of all the other management system elements.

MONITOR AND MANAGE PROCESS PERFORMANCE—LISTEN TO THE VOICE OF THE PROCESS

How can you tell how well a work process is performing? By collecting and interpreting the right data while it is operating. (See the discussion on measurement in Chapter 7, "Process Quality Assurance.") The data you collect should help you (as a minimum) monitor and manage those factors or variables in the work process that directly cause customer satisfaction. Combined with your knowledge of variation and the appropriate statistical tool, quality results data should tell you when, or whether, changes to the work process are desirable, or, conversely, would be counterproductive. (See the discussion on variation in Chapter 7.) Because the right quality results data can tell us so much, we refer to it as "the voice of the process."

OBTAIN FEEDBACK

A second reason why quality results data is so important is that it serves as a primary source of feedback on how well the overall QMS is working. In particular, quality results data closes the measurement system loop that consists of key linkages between your Total Customer Satisfaction, Strategic Quality Planning, Process Quality Assurance, and Quality Results processes. (See the discussion under "What happens to the measurement data we collect?" later in this chapter.)

HOW DO QUALITY RESULTS PROCESSES DIFFER FROM TOTAL CUSTOMER SATISFACTION PROCESSES?

Since both Total Customer Satisfaction and Quality Results processes collect data, managers often wonder what the distinctions are between the two. One main distinction is that Total Customer Satisfaction processes provide external feedback, whereas Quality Results processes provide internal feedback. In the former case, you are collecting data from your customers *after* they have received the outputs (products or services) from your work processes. In the latter case, *you* see quality results data hopefully while the output is being produced and certainly before it goes to a customer (unless you are providing a pure service where at least some of the output is produced and perceived by the customer at the same time, for example, a barber giving a haircut).

In a sense, Quality Results processes represent a bridge between your QMS and your customers. Quality Results data tell you "if you are doing *things right,*" and Total Customer Satisfaction data tell you from your customer's perspective "if you have *done the right things.*" This "correlation between quality and customer satisfaction is a critical management tool."[1]

If well designed, your Quality Results processes will act as an early warning system for customer satisfaction. They will use measures to collect data real-time (while the process is operating) that may be used to predict how

---

[1]Section 6.1, Malcolm Baldrige National Quality Award Application.

customers will perceive the outputs once they receive them. In a mature, well-designed QMS, your customer satisfaction data will *confirm* what your quality results data predicts, unless, of course, customer expectations have changed.

How can you make sure that your quality results processes help predict customer satisfaction and provide feedback that you can use to better monitor and manage work process performance? By making sure that your collection processes collect the right data, provide that data in an easy-to-understand manner, do so in a timely fashion, and distribute the data so that it may be used in other key parts of the QMS.

WHAT IS THE RIGHT DATA TO COLLECT?

This question should be answered as part of your Process Quality Assurance processes. If you recall, you create process performance measures in that part of the QMS. Generally speaking, the right data to collect is:

1. Tied directly to customer expectations and the resultant process performance targets.

2. Useful as a real-time indicator of critical process variables that impact customer satisfaction.

3. Focused on how well the process is performing vs. simply how productive it is. (See the discussion on measurement in Chapter 7, "Process Quality Assurance.")

EASY TO UNDERSTAND

As you collect quality results data, you should compile, organize, and communicate in an easy-to-understand manner. It should clearly indicate the status of those relevant aspects of process performance that you decided to measure as part of your Process Quality Assurance processes. Ideally, you should display the data where it can be viewed by everyone working in the process so that they can see at a glance how well the process is performing.

For example, Federal Express as a corporation tracks several key measures of performance. Each of these measures is tied directly to a key customer requirement. Federal Express promises on-time delivery and thus measures many aspects of on-time performance to ensure that it meets its delivery goal. The company reviews the previous day's results for each service measure with each station manager every morning via a satellite TV broadcast.

TIMELY

For quality results data to be helpful in monitoring and managing a work process while it is operating, it must be readily available without delay. This is a major challenge for many organizations. They simply do not have the capability, even manually, to do this. Leading firms recognize the competitive advantage that the right data, available in a timely fashion, provide. John Martin, CEO of Taco Bell, is able to pull up real-time sales updates from any of Taco Bell's 3,000 company stores in 15 minutes.[2] The firm's information systems capability is one reason that Taco Bell's revenues grew from $1.6 billion to approximately $4.3 billion from 1988 to 1994.

WHAT HAPPENS TO THE QUALITY RESULTS DATA YOU COLLECT?

For quality results data to contribute to managing the business, as in the preceeding example, it must feed into and become part of these other elements of your QMS:

Strategic Quality Planning

1. Establish Improvement
    Objectives

Management By Fact

3. Make Decisions

**Use Quality
Results Data
to:**

2. Evaluate Improvements/
    Monitor Performance

Process Quality Assurance

4. Provide Feedback

**Human Resources
Effectiveness**

FIGURE 9.1   Four Ways to Use Quality Results Data

---

[2]*Forbes,* August 29, 1994.

As part of your Strategic Quality Planning processes, you use Quality Results data to establish improvement objectives. Your Process Quality Assurance processes use Quality Results data to evaluate improvement project results and for real-time monitoring of process performance. One of the most important ways to use Quality Results data is to support the decisions you make as part of your Management by Fact processes. As a manager, it is important for you to demonstrate that Quality Results data will be used to make decisions. A famous quote, attributed to the Milliken Company, goes like this: "In God we trust; everyone else brings data." Make sure your data is a regular part of management presentations, updates, and meetings. Bob Galvin, former chairman of Motorola, was famous for starting his weekly staff meetings by first reviewing the quality data, and then leaving the room while other issues such as financial data were discussed. Finally, you should use the data you collect from your Quality Results processes to provide feedback as part of your Human Resources Effectiveness processes.

One final note on quality results data: Avoid collecting data that you do not intend to use. Author and consultant James Harrington states: "It's clear that if you can't measure an activity, you can't improve it. But measurement without feedback is worthless because you've expended the appraisal effort but have not provided the individual with the opportunity to improve."[3]

The following examples illustrate how managers at three different organizations apply quality results practices. As you read each example, ask yourself, "What stage of implementation best characterizes this organization," and, "Are the actions this manager takes appropriate?"

EXAMPLE 1

Carol manages a district sales force for a firm in the medical supplies industry. The company distributes a wide range of supplies made by major manufacturers such as Johnson & Johnson. To better serve their joint customers, Carol's firm has been working on improvement teams with several of the manufacturer's personnel. One of the teams has been studying the order-fulfillment process for several months. This project is one of the first to involve the sales organization. Carol was asked to join the team because of her sales, customer, and improvement process knowledge. She also is widely respected within her own organization. As part of the improvement project, she will be documenting and analyzing the firm's existing order-fulfillment process performance. She expects that the project will help improve the timeliness,

[3]*Business Process Improvement,* H. James Harrington, McGraw-Hill, 1991 p. 184.

availability, or usefulness of the order-fulfillment process performance data currently being collected.

EXAMPLE 2

The Federal Express corporation (the first service organization to win a Baldrige Award a few years ago) tracks several key measures of performance. Each of these measures is tied directly to a key customer requirement. Federal Express promises on-time delivery and thus measures many aspects of on-time performance to ensure that they meet their delivery goal. The company reviews the previous days results for each service measure with each station manager every morning via a satellite TV broadcast.

EXAMPLE 3

An article in the August 8, 1994, issue of *Business Week* (primarily explaining return on quality) stated that Robert Allen, CEO of AT&T, receives a quarterly report from each of the company's 53 business units that spells out quality improvements and their subsequent financial impact. The article also states that based on AT&T's experience, when customers perceive improved quality, it shows up in better financial results three months later. "This is the most important thing that AT&T has ever done," Allen told a meeting of top managers the day before his June board presentation. The article goes on to say that to win approval from AT&T's top management these days, proponents of any new quality initiative must first demonstrate that the effort will yield at least a 30 percent drop in defects and a 10 percent return on investment.

How did you do? Check your understanding of quality results actions based on the following discussion.

EXAMPLE 1

Carol is working on one of the first improvement projects in the sales function; this usually means the organization is in Stage 2 of implementation. Her action of improving the timeliness, availability, or usefulness of the order-fulfillment process-performance data currently being collected is appropriate for this stage.

EXAMPLE 2

Federal Express is in Stage 3 of implementation. The company has been working on quality improvements for several years now. The daily reviews of results data with each station manager is a good example of institutionalizing the links between your Total Customer Satisfaction, Strategic Quality Planning, Process Quality Assurance, Human Resources Effectiveness, Management by Fact, and Quality Results processes.

EXAMPLE 3

This example from AT&T also describes an organization in Stage 3 of implementation. The quarterly reports of quality results data from 53 business units show how AT&T has institutionalized links between its Total Customer Satisfaction, Strategic Quality Planning, Process Quality Assurance, Human Resources Effectiveness, Management by Fact, and Quality Results processes. It also shows widespread use of the quality results data it collects.

SUMMARY

- Quality Results processes collect, compile, organize, and communicate work process performance data.

- The primary use of this data is feedback.

- Quality Results data directly supports these other QMS elements:
  — Strategic Quality Planning
  — Product/Service Quality Assurance
  — Management by Fact
  — Human Resources Effectiveness

- Combined with customer expectations data, quality results data helps you establish improvement objectives as part of your Strategic Quality Planning processes.

- Real-time quality results data allows you to monitor and manage process performance as part of your Process Quality Assurance processes; the data also tell you whether improvement actions have been successful.

- Quality results data are essential if you are to Manage by Fact (i.e., "Without data, you're just someone else with an opinion").

- Quality results data should be incorporated into the feedback provided as part of your Human Resources Effectiveness processes.

- Quality results *data* come from the measures established as part of your Process Quality Assurance processes. (See the discussion on measurement in Chapter 7, "Process Quality Assurance.")

QUALITY RESULTS ACTIONS CHECKLIST

Stage 1

1. Align the balance of consequences so that they reinforce the collection of quality results data.
2. Begin collecting quality results data.
3. Identify the quality results data currently available.
4. Identify the methods currently used to collect quality results data.
5. Identify the methods currently used to compile, organize, and communicate quality results data.
6. Assess what happens to the quality results data you currently collect.
   — How is it used?
   — Who uses it?
7. Assess the timeliness, availability, or usefulness of the quality results data you currently collect.
8. Evaluate the links between your Total Customer Satisfaction, Strategic Quality Planning, Process Quality Assurance, Human Resources Effectiveness, Management by Fact, and Quality Results processes.
9. Establish links between your Total Customer Satisfaction, Strategic Quality Planning, Process Quality Assurance, Human Resources Effectiveness, Management by Fact, and Quality Results processes.
10. Coach employees in collecting and analyzing quality results data.

Stage 2

1. Improve the timeliness, availability, or usefulness of the quality results data you currently collect.

2. Increase the use of quality results data you collect.

3. Improve the methods used to collect quality results data.

4. Improve the methods currently used to compile, organize, and communicate quality results data.

5. Strengthen or improve the links between your Total Customer Satisfaction, Strategic Quality Planning, Process Quality Assurance, Human Resources Effectiveness, Management by Fact, and Quality Results processes.

6. Use quality results data to calculate the return on quality.

7. Design or modify your work processes and jobs so that they directly provide quality results data to associates while the process is operating (i.e., create a built-in feedback capability).

Stage 3

1. Ensure widespread use of the quality results data you collect.

2. Automate the methods used to collect quality results data.

3. Automate the methods used to compile, organize, and communicate quality results data.

4. If you haven't already done so, integrate your quality results processes and data into your company's overall information system capability.

5. Incorporate best-in-class benchmarks into your quality results data.

6. Institutionalize the links between your Total Customer Satisfaction, Strategic Quality Planning, Process Quality Assurance, Human Resources Effectiveness, Management by Fact, and Quality Results processes.

CHAPTER TEN

HUMAN RESOURCES EFFECTIVENESS

Everything that enlarges the sphere of human powers, that shows man he can do what he thought he could not do, is valuable.
—Samuel Johnson

EXECUTIVE OVERVIEW

- There is a wide variety of processes that impact the effectiveness of your associates.

- Most managers tend to rely on a relatively narrow subset of the processes available.

- One of the most powerful processes managers may use to increase People Effectiveness is to align the balance of consequences the organization provides with the personal values of each associate.

WHAT IS HUMAN RESOURCES EFFECTIVENESS?

Human Resources Effectiveness processes provide each associate with the environment, know-how, and support systems to better realize their potential at work. Ideally these processes work together collectively to assure a good fit between what associates want and are well suited to do, and what the organization's work processes require in order to satisfy internal and external customers.

Here is a list of some of the more well-known human resources effectiveness processes:

- Staffing
- Orientation

- Training
- Compensation/Benefits
- Recognition
- Performance Appraisal
- Feedback
- Career Planning
- Promotion/Succession Planning
- Professional Development

THE FEWER THE BETTER?

It seems like many managers tend to rely almost exclusively on one or two processes to increase the effectiveness of their associates. Perhaps this is due to a belief that individual managers have little direct influence over many of the Human Resources Effectiveness processes at their disposal.

Take another look at the list of Human Resources Effectiveness processes above. How many do *you* feel you can directly influence?

Interestingly enough, even though managers often feel like they have little direct influence over the processes listed above, associates often believe otherwise. In fact, in the eyes of your associates, *you* are the organization's people policies and procedures, as well as all the other elements of the quality management system (QMS).

Fortunately, there is a way out of this apparent dilemma. Since your associates already believe that what *you* say and do sends a direct message about what the organization values, expects, and rewards, you may as well make this situation work for you rather than against you. You can do this by aligning the balance of consequences that your associates experience so that the organization (you) reinforces the personal values of each associate.

BALANCE OF CONSEQUENCES

Why aren't more companies successful at implementing quality improvement? More importantly, if most individuals believe that improving quality is the right thing to do, then why aren't more people volunteering to work on improvement projects or doing those things that make quality happen? We contend the answer, perhaps more than any other single reason, is that the current balance of consequences people experience in their workplace *does not favor quality improvement*. Stated another way, the "momentum" in the organization strongly favors the status quo, or business as usual.

Consider the following situation:

You are the manager of an assembly area in a plant that makes telecommunication products. You have 15 people that report to you, all of whom are skilled technicians with varying years of service. Your company has established a continuous improvement process and all your people have attended training in how the improvement process works. You've also attended the training, and buy in to the fact that unless the company makes continuous improvement a way of life, competitive forces in the industry will continue to cause downsizing, consolidation, and all of the other things that put your job at risk. Besides, you personally believe that increasing customer satisfaction is the right thing to do, and you want to do your part. Accordingly, you study customer satisfaction data and decide to start an improvement project to reduce the number of defects associated with the assemblies your area produces. You create a team made up of three people who work on the assembly process: George Johnson, Mary Adams, and Chris Cobb. You call the three of them together, review the customer satisfaction data with them, and challenge them to use the company's improvement process to find and eliminate the root causes of the defects. You explain to the team that this is an important project, not just because it is the first one to be done in your area, but because the assemblies made here are vital to the overall profitability of one of the firm's major product lines. You establish a regular weekly schedule to meet with the team to discuss progress or otherwise work with them as they see fit. Near the end of the meeting, you ask, "Are there any questions about this project?" The three of them look at each other, then at you. Finally, George says, "Frank over in Building C has been working on a quality project. He says that it puts a lot of pressure on him because he's still responsible for his regular work. Is that the case here too?" Then Chris speaks up. "As you know, my wife just had a baby, our first. We both work, and we've agreed to share the child-rearing responsibilities evenly. If this project means overtime, I can't do it, even if it's at time-and-a-half." Mary then says, "I like the idea of trying to make things better, I really do. But it's been over six months since I attended that quality training course you sent me to. I'm afraid I'm a little rusty on what they covered." "Me too," the others chime in. "Is there anyone we can turn to for help if we have any questions?"

"Hold on," you say. "Let's take these one at a time. Now, about the work load . . ."

Is this project off to a good start? How would you proceed at this point? Would you have done anything different prior to the meeting? During the meeting?

Believe it or not, this project may be off to a better start than many with which we're familiar. Why? Because the team members are at least voicing their concerns. We know of many cases where the manager asks, "Are there any questions?" and no one says anything. Sometimes the projects just drag on until the manager "discovers" that there is a problem.

But what's wrong with this picture? Are these typical concerns being voiced by the improvement project team members, or is this manager stuck with "uncooperative" people?

To answer this question, it helps to look at the situation more closely to see if we can figure out what is really going on and identify what you as a manager can do to get your own improvement projects off to a great start.

First, we'll present a framework to help you analyze performance. Then we'll show how you can use that framework to increase the likelihood that the desired performance will actually occur.

Our experience is that most managers have never learned the framework we are about to present, yet it is based on more than 50 years of experimental and applied research.[1] Simply put, it works! The information and underlying concepts presented here are further described in *Performance Management,* by Aubrey Daniels. Our intent is to show how these concepts may be applied by managers to help implement a QMS. All credit for development of the "ABC model of behavior change" belongs to Daniels.

THE ABC'S OF PERFORMANCE

Performance is a combination of what people do (their behavior) and the results they achieve (their accomplishments or outputs). There are two main ways to change behavior. You can try to influence behavior *before* it happens or *after* it occurs. The things you use to try to influence what people do before they do it are called "antecedents." When you try to influence behavior *after* it takes place, you provide "consequences."

By recognizing and understanding the relationships among these three elements (antecedents, behavior, and consequences), you can analyze the effect each has on performance and adjust one or more elements to help produce the desired results.

The pattern you wish to recognize is this: An antecedent prompts a behavior, which is followed by a consequence. An easy way to remember the three elements is through the first initials of each, ABC. ABC analysis, as it is called, is the process of systematically examining all the antecedents and

---

[1] Aubrey C. Daniels, *Performance Management* (Performance Management Publications, 1989) p. 8.

consequences associated with a given behavior or performance in order to better understand why people do what they do (Daniels, p. 37).

Before we show you how to do an ABC analysis, let's better define each of the three elements.

ANTECEDENTS

Earlier, we said that antecedents are the things you use to try to influence what people do before they do it. Since the term antecedent sounds more like a mouthwash, or maybe toothpaste, to many people, here is a definition that you can work with. According to Daniels, an antecedent is "a person, place, thing, or event coming before a behavior that encourages you to perform that behavior." Does that help? Okay, try this. An antecedent is anything that serves as a signal or cue to a person and that communicates information about behavior and its consequences (See Table 10.1).

TABLE 10.1   Six Most Common Types of Antecedents Used to Prompt Work Behavior

Type of Antecedent	Example(s)
Job aids	Instructions, directions, signs, labels diagrams, checklists, color codes
Training	Workshops, textbooks, videos
Tools and materials	Any resource to help someone do their job
Policies and procedures	Job descriptions, rules, guidelines
Work environment	Workplace conditions such as temperature, housekeeping, noise levels
Meetings	All types

Source: Aubrey C. Daniels; *Performance Management* (Performance Management Publications; 1989) p. 19.

If you're like most managers, you are very familiar with all of these antecedents, though you probably never heard them referred to as antecedents.

Here are some things you may **not** know about antecedents:

- Antecedents get a behavior started, but only consequences maintain behavior (Daniels, p. 17).

- In the workplace, if an antecedent gets a behavior to occur **one** time, it has done its job. That is all that can be expected of it. Whether the behavior occurs again will be determined for the most part by the consequences associated with it (Daniels, p. 17).

- Data collected by Aubrey Daniels shows that, on average, managers spend approximately 85 percent of their time on antecedents— telling people what to do, figuring out what to tell them to do, or figuring out what to do because people didn't do what they told them to do (p. 17).

- Managers spend too much time on antecedents and not enough on providing consequences, yet consequences are the single most effective tool a manager has for increasing employee performance and improving morale, according to Daniels (p. 23).

CONSEQUENCES

Consequences are events that follow behaviors and change the probability that they will recur in the future (Daniels, p. 23). Here's a quick example. You put the correct change in a soft drink machine and press the button for the soft drink you desire (a behavior or action). If all is well, your soft drink comes out (the consequence). What happens if the machine takes your money? (A different consequence.) Will you try again? Go to a different machine? When the consequence changes, behavior changes. An extremely important set of consequences comes from what you as a manager do, say, provide, or withhold as perceived (experienced) by each of your associates.

Here are a few key points to remember regarding consequences:

- Every behavior has a consequence.

- Consequences affect performance whether or not they are managed.

- Every individual experiences consequences differently; different people often respond differently to the same consequence.

- There are two basic type of consequences—those that increase performance (positive) and those that decrease performance (negative).

As stated in Chapter 2, effective managers analyze and align the "balance" of consequences (the net affect of positive and negative consequences that a person experiences after a given behavior or action) so that they more strongly encourage or reinforce quality improvement.

ANALYZING CONSEQUENCES (NOT ALL CONSEQUENCES ARE CREATED EQUAL)

There are three main characteristics of consequences that determine the influence they have on behavior:

- Whether the *performer* views the consequence as positive or negative.

- The timing of the consequence.

- The certainty of the consequence.

POSITIVE OR NEGATIVE—IT'S IN THE EYE OF THE BEHOLDER

Suppose someone that works for you makes a suggestion during one of your regular staff meetings. You think its a good idea, and rather than communicating it further up the line yourself, see it as a great opportunity to provide the person with some exposure and recognition, so you say, "That's a great suggestion. I'd like you to prepare a 15-minute presentation that you will make yourself to the director during our operations review meeting tomorrow."

Will the consequence of preparing and making a 15-minute presentation for the director (within 24 hours) be viewed as positive or negative *by this associate?* The answer depends on what this *associate* feels about the "developmental opportunity" you've just provided.

What about the other people attending the same meeting? They have just seen the behavior "making a suggestion" and heard the consequences. How will each of them react the next time *they* think about making a suggestion? Again, it depends on each individual's likes and dislikes. The point is that what one person considers a positive experience, another may view as negative. To make effective use of consequences as a manager, you must learn enough about each person that works with you so that you can provide the right types of experiences (those that will be viewed positively by that employee) when you want to. There are few things more frustrating (to both you and your associates) than consequences that have the opposite effect of what you intended.

What difference does it make? Consequences that are viewed by the performer as positive increase the likelihood that they will repeat the behavior again. Those that are viewed as negative decrease the chances that the behavior will occur again.

This certainly isn't rocket science. All we are saying is that if you take an action and the result is getting something you want, or avoiding something you don't want, the chances are pretty good you might take that action again.

On the other hand, if *for the same action* you get something you don't want, or don't get something you do want, chances are you'll be less likely to repeat the action.

CONSEQUENCE TIMING—NOW IS
DEFINITELY BETTER THAN MAYBE LATER

Consequences have their greatest impact on behavior when they are immediate. At work, the most immediate consequences are those that occur while the person is working (Daniels, p. 28). Here's an everyday example (perhaps for someone you know). Suppose the posted speed limit on the main thoroughfare between two very busy roads is 35 m.p.h. The thoroughfare winds its way lazily through a combination of residential and commercial property for about three miles. You have to take this road during your daily commute or else it doubles your commute time. You've also discovered that there is a short 10-minute "window" during which if you make it to one of the busier roads, it cuts fifteen minutes off your time. If you miss the window, it adds 20 minutes to your daily commute. One day you are passed by the car following you, which is driving above the speed limit! You watch carefully and *nothing happens.* Are they likely to do this again? Would you ever try this? This is an example of delayed, or **future,** consequences.

Now suppose it's another day. Once again the car behind you passes, and sure enough continues to drive above the speed limit. This time, seemingly from out of nowhere, a siren sounds, and our luckless friend is about to experience **immediate** consequences from his actions.

CONSEQUENCE CERTAINTY—ISN'T IT NICE
TO KNOW THAT THERE ARE SOME THINGS
IN LIFE YOU CAN COUNT ON?

What do death and taxes have in common? According to many, they are the only two things in life you can count on. When you can always count on the fact that a given consequence will follow a behavior (i.e., that consequence is

certain to occur), then it will have a great deal of impact on the behavior it is associated with. Generally speaking, the more certain a consequence is, the greater the impact it will have on the behavior it follows.

Using the driving example we described earlier, let's see how these three characteristics of consequences work together to influence behavior. This time, though, we'll use the ABC framework and we will classify each of the consequences present for the driver of the car as positive or negative, immediate or future, and certain or uncertain.

First let's look at the behavior of driving *faster* than the posted speed limit. See Figure 10.1.

Now, for the *same* driver, let's see what happens if we describe the *desired* behavior of driving at the posted speed limit. See Figure 10.2.

Look closely at the consequences for both behaviors. A consequence that is classified as positive, immediate, and certain (PIC), or one that is negative, immediate, and certain (NIC), has the greatest impact on any behavior. PICs *reinforce* the behavior; NICs stop or decrease behavior. In our speeding example, the consequences for this driver strongly reinforce the behavior of driving faster than the posted speed limit, and they work to decrease the behavior of driving at the posted speed limit.

Now you have a scientific explanation of why some people speed. The next time someone you know is stopped for speeding, they can explain that they were speeding because they suffer from a PIC consequence deficiency.

WHAT'S A MANAGER TO DO?

The preferred approach (based on research and results) is to focus your attention on providing more PICs when you want to reinforce or encourage a particular behavior. You should also consider adding new antecedents as long as you can closely associate them with PICs for the behavior you wish to strengthen.

According to Daniels, an ABC analysis is very helpful when facing any situation involving resistance to change, or when planning for a new performance. It will usually show:

- Many things currently present in the work environment (antecedents) *do not prompt* the desired behavior.

- The balance of consequences (the mix of P/Ns, I/Fs, and C/Us) currently present work to decrease or stop the desired behavior.

Let's revisit the earlier staff meeting now that we can see things differently and do an ABC analysis for one of the attendees (remember, consequences are experienced differently for *each* individual).

Name: Car Driver

Antecedents	Consequences	Positive or Negative	Immediate or Future	Certain or Uncertain
• Speed limit sign shows 35 m.p.h.	• Gets home sooner	P	I	C
• Sees others driving faster than 35 m.p.h.	• Avoids traffic delays down the road	P	I	C
• Has previously driven faster than 35 m.p.h.	• May get ticket	N	F	U
• Does not see police cars in the area				

Undesired Behavior:

Driving faster than the posted speed limit.

Instructions:

1. Complete two worksheets for each individual (start with the *undesired* behavior and then complete a worksheet for the **desired** behavior).
2. Write the person's name and the **desired** behavior for whom you are doing this analysis.
3. List all the possible antecedents that come to mind associated with the **desired** behavior.
4. List all the possible consequences that come to mind associated with the **desired** behavior.
5. Review the list of consequences and mark out those that are relevant to the company rather than the performer.
6. Classify the remaining consequences as PN, IF, or CU.
7. Review the antecedents and consequences currently in place. Does the balance (mix) reinforce what you want it to? If not, align the balance of consequences; add Ps & Is and antecedents for the **desired** behavior.

FIGURE 10.1   Balance of Consequences Worksheet

Name: Car Driver

Antecedents		Consequences	Positive or Negative	Immediate or Future	Certain or Uncertain
• Speed limit sign shows 35 m.p.h.		• Gets home later	N	I	C
• Sees others driving the legal limit		• Gets caught in delays	N	I	C
• Does not see police cars in the area		• Avoids ticket	P	I	C

Desired Behavior:

Driving at the posted speed limit.

Instructions:

1. Complete two worksheets for each individual (start with the *undesired* behavior and then complete a worksheet for the **desired** behavior).
2. Write the person's name and the **desired** behavior for whom you are doing this analysis.
3. List all the possible antecedents that come to mind associated with the **desired** behavior.
4. List all the possible consequences that come to mind associated with the **desired** behavior.
5. Review the list of consequences and mark out those that are relevant to the company rather than the performer.
6. Classify the remaining consequences as PN, IF, or CU.
7. Review the antecedents and consequences currently in place. Does the balance (mix) reinforce what you want it to? If not, align the balance of consequences; add Ps & Is and antecedents for the **desired** behavior.

FIGURE 10.2 Balance of Consequences Worksheet

Figures 10.3 and 10.4 display the ABC analysis for Mary Adams. With this and similar analyses in hand for Chris and George, we can better answer the questions posed at the outset. We see that the balance of consequences clearly does not favor any of these team members taking on this assignment. In fact, the number one consequence mentioned most by attendees at our workshops "is quality improvement adds to my current work load."

What should you do if you want any of these three people to more readily work on the project? Add positive, immediate, certain consequences (PICs) and antecedents for participating, and change the specific mix for each individual so that it reinforces working on the project. This is another reason making improvement projects voluntary is helpful, and then visibly recognizing and rewarding those who do volunteer is a successful strategy.

It has been a while since we referred to our fitness analogy, but if you really want to improve your level of fitness, do an ABC analysis on some of the key behaviors involved. Figures 10.5 and 10.6 show a completed balance of consequences worksheet for the behaviors of eating sweets vs. not eating sweets.

Now that you have a better understanding of ABC analysis, it should be clearer to you that a key reason so many people fail to improve their level of fitness is because the balance of consequences they currently experience does not reinforce the desired behaviors associated with becoming fit. Unfortunately, the same *holds true for quality improvement.* If you want to make quality happen in your part of the business, you must align the balance of consequences so that they favor actions that improve quality for each of your associates.

Take another look at the common Human Resources Effectiveness processes listed at the beginning of this chapter. Which of the three ABCs (elements) of performance best describes these processes? *Hint:* Look at Table 10.1 (Go to the next MENSA meeting if you concluded that almost all the processes listed are antecedents!)

Now think back to some of the key points to remember about antecedents (see list on pages 127–128). If you really want to increase the effectiveness of these processes, look for ways to associate (link) these antecedents with a specific behavior and then provide positive, immediate, and certain consequences for that behavior.

Here's one example. One of the most common and frequently used antecedents is training. If you want to get more value from any training, link it to an aspect of desired job performance and then provide positive, immediate, certain consequences for your associates when they return from the training and demonstrate the desired performance. Remember our three assembly technicians? The manager sent all three to the company's continuous improvement training (antecedent) but apparently did not link that event with

Name: Mary Adams

Antecedents

- Hears peers talk about how much time it takes

- Concerned about not knowing how to follow the continuous improvement process

- Sees others work late to catch up with normal work load

- No one else in this area is doing it

Undesired Behavior:

Not working on improvement project.

Consequences	Positive or Negative	Immediate or Future	Certain or Uncertain
• Avoids getting behind on regular work	P	I	C
• Avoids looking ignorant in front of peers	P	I	C
• Might negatively affect next performance appraisal	N	F	U
• Might get a lecture on "being a team player"	N	F	U

Instructions:

1. Complete two worksheets for each individual (start with the *undesired* behavior and then complete a worksheet for the **desired** behavior).
2. Write the person's name and the **desired** behavior for whom you are doing this analysis.
3. List all the possible antecedents that come to mind associated with the **desired** behavior.
4. List all the possible consequences that come to mind associated with the **desired** behavior.
5. Review the list of consequences and mark out those that are relevant to the company rather than the performer.
6. Classify the remaining consequences as PN, IF, or CU.
7. Review the antecedents and consequences currently in place. Does the balance (mix) reinforce what you want it to? If not, align the balance of consequences; add Ps & Is and antecedents for the **desired** behavior.

FIGURE 10.3   Balance of Consequences Worksheet

Name: Mary Adams

Antecedents	Desired Behavior: **Working on improvement project.**	Consequences	Positive or Negative	Immediate or Future	Certain or Uncertain
• Regular weekly meetings with the boss		• Attending weekly meetings will take time away from other things	N	I	C
• Training on the continuous improvement process		• May have to attend more train-ing, which will take more time	N	I	C
• High-visibility project		• Working on this project may be risky	N	I	C
• Encouragement from management		• People that don't know me will be judging my work on this project	N	F	C
• Likes learning		• Project may make things better	P	F	U
		• May get recognition	P	F	U
		• May learn new skills which could lead to better job	P	F	U

Instructions:

1. Complete two worksheets for each individual (start with the *undesired* behavior and then complete a worksheet for the **desired** behavior).
2. Write the person's name and the **desired** behavior for whom you are doing this analysis.
3. List all the possible antecedents that come to mind associated with the **desired** behavior.
4. List all the possible consequences that come to mind associated with the **desired** behavior.
5. Review the list of consequences and mark out those that are relevant to the company rather than the performer.
6. Classify the remaining consequences as PN, IF, or CU.
7. Review the antecedents and consequences currently in place. Does the balance (mix) reinforce what you want it to? If not, align the balance of consequences; add Ps & Is and antecedents for the **desired** behavior.

FIGURE 10.4   Balance of Consequences Worksheet

Name: Fred

Antecedents

- Hears peers talk about how good it tastes
- Others at the table are eating it
- Sees how good it looks

Undesired Behavior:

Eating sweets.

Consequences	Positive or Negative	Immediate or Future	Certain or Uncertain
• Tastes good	P	I	C
• Makes me feel like one of the gang having fun	P	I	C
• Might gain weight	N	F	U

Instructions:

1. Complete two worksheets for each individual (start with the *undesired* behavior and then complete a worksheet for the **desired** behavior).
2. Write the person's name and the **desired** behavior for whom you are doing this analysis.
3. List all the possible antecedents that come to mind associated with the **desired** behavior.
4. List all the possible consequences that come to mind associated with the **desired** behavior.
5. Review the list of consequences and mark out those that are relevant to the company rather than the performer.
6. Classify the remaining consequences as PN, IF, or CU.
7. Review the antecedents and consequences currently in place. Does the balance (mix) reinforce what you want it to? If not, align the balance of consequences; add Ps & Is and antecedents for the **desired** behavior.

FIGURE 10.5   Balance of Consequences Worksheet

137

Name: Fred

Antecedents
• Hears peers jokingly say, "You're no fun"
• Others at the table are eating sweets
• Calculates fat content

Desired Behavior:

Not eating sweets.

Consequences	Positive or Negative	Immediate or Future	Certain or Uncertain
• Makes me feel like I'm spoiling the gang's fun	N	I	C
• Makes me feel left out	N	I	C
• Avoids gaining weight	P	F	U

Instructions:

1. Complete two worksheets for each individual (start with the *undesired* behavior and then complete a worksheet for the **desired** behavior).
2. Write the person's name and the **desired** behavior for whom you are doing this analysis.
3. List all the possible antecedents that come to mind associated with the **desired** behavior.
4. List all the possible consequences that come to mind associated with the **desired** behavior.
5. Review the list of consequences and mark out those that are relevant to the company rather than the performer.
6. Classify the remaining consequences as PN, IF, or CU.
7. Review the antecedents and consequences currently in place. Does the balance (mix) reinforce what you want it to? If not, align the balance of consequences; add Ps & Is and antecedents for the **desired** behavior.

FIGURE 10.6   Balance of Consequences Worksheet

138

any specific behavior and provided weak consequences at best (it had been over six months since anyone attended, and nothing happened either positive or negative until this meeting). If our manager had done an ABC analysis, he would have known to schedule the training so that it would be completed just prior to starting the improvement project. Then, to link the training to desired performance, he would have sat down with each technician individually prior to the training, explained the upcoming improvement project, and emphasized how the training would help prepare the technician to successfully participate on the project. The manager would have anticipated at least some of the concerns of the technicians (as a result of the ABC analysis) and have been able to address them before sending the associate to the training. When the technicians returned from the training, the manager would have provided one or more positive, immediate, consequences (tailored to each technician) in order to reinforce (recognize, encourage, reward) the effort that the technician had demonstrated. For instance, perhaps the manager could let the technician leave work early with pay, acknowledge the technician's actions to prepare for the upcoming project during the next staff meeting, write a personal note thanking the technician, etc. The key is to provide a consequence that *each* technician will personally experience as positive, immediate, and certain.

STACKING THE DECK

Early in this book (see Chapter 2, "What Do Successful Managers Do, Anyway?"), we listed the following three key principles that guide the actions of effective managers of quality improvement:

- Integrate the quality message into daily communication.

- Visibly spend time taking actions to improve quality daily.

- Align the balance of consequences to reinforce quality improvement.

Now that you have come this far, you may as well learn the secret to managerial success. Really effective managers take actions that combine all three principles at once! For example, they verbally recognize and praise an associate's quality-related action at the beginning of each day in front of the whole group.

The more often you can do this in your role as manager, the more you stack the deck so that quality improves—and everyone wins!

The following examples illustrate how managers at different organizations apply Human Resources Effectiveness practices. As you read each example, ask yourself, What stage of implementation best characterizes this organization? Are the action(s) this manager takes appropriate?

EXAMPLE 1

Dale manages a group of salespeople for a major computer manufacturer. The firm is experiencing intense competition in its primary market of workstations. Long known for technology, the newly hired CEO (from outside) has announced plans to focus the firm's well-known sales force on increasing customer satisfaction by eliminating straight commission and replacing it with 50 percent salary, 50 percent bonus. The bonus is based on achievement of minimum profitability and customer satisfaction levels. The firm has recently initiated large-scale process improvement teams, but no one from sales has been involved to date. The rumor mill is saying that customers want features such as built-in networking and telecommunications capability. Dale knows these won't be available until the next generation of the company's workstations is announced.

EXAMPLE 2

Kathy manages a group of salespeople for a major computer manufacturer. The firm is experiencing intense competition in its primary market of workstations. Long known for technology, the newly hired CEO (from outside) has announced plans to focus the firm's well-known sales force on increasing customer satisfaction by eliminating straight commission and replacing it with 50 percent salary, 50 percent bonus. The bonus is based on achievement of minimum profitability and customer satisfaction levels. The firm has successfully completed many large-scale process improvement projects over the last four years, including several involving all aspects of the sales function. As a result of one of these projects, the company is ready to announce, support, and service a full line of workstations with features their customers have asked for, such as built-in networking and telecommunications capability. Beta test results with key customers have resulted in advance commitments for the new products. Kathy can't wait to let her sales force know about the new compensation policy.

EXAMPLE 3

John is planning to start an improvement project in the area he manages. His company has collected and analyzed customer data and used it to identify 10 key processes that impact customer satisfaction, though process owners have not yet been identified. The firm established overall cycle time and defect-reduction improvement objectives for those processes and has briefed all the managers involved (John is one of those managers). John has just attended training in continuous improvement as well as a course titled, "How to Effectively Sponsor an Improvement Project." As John reviews the data the firm collected and the resultant improvement goals, his thoughts turn to improvement team member selection. He knows that its not enough to simply appoint a team, send them to training, and let the project take care of itself. He wants to get the project off to a good start, so he schedules the training so that it will be completed just prior to starting the improvement project. Then, he plans to sit down with each team member individually prior to the training, explain the upcoming improvement project, and point out how the training will help prepare the associate to successfully participate in the project. He decides to anticipate and address any concerns that each team member may have by identifying how *they* will view working on the project based on his understanding of what makes each tick. He also plans a series of actions that he will take for each team member to encourage and reward the effort that each demonstrates upon returning from training.

You can check your understanding of People Effectiveness practices by reading the following discussion.

EXAMPLE 1

Poor Dale. He's in for a tough time. Why? Because his firm is trying to use a Stage 3 action (changing compensation to reward profitability and customer satisfaction) while the company is early in Stage 1 of implementing a QMS. Now compare this example with Example 2.

EXAMPLE 2

Kathy is about to be viewed as a hero by her sales associates. Her firm is adopting the exact same compensation policy, but the company is ready for it. This company is in Stage 3 of its QMS implementation. It has had several successful large-scale process improvement projects, some of which have involved sales. The company has also worked closely with customers and other parts of the business to offer and support the features that customers want right now. It's going to be a fun year for Kathy once the sales force learns about the new compensation changes.

EXAMPLE 3

John is beginning to identify the current balance of consequences for each prospective improvement project team member. This is an appropriate action since his company is in Stage 1 of implementation.

SUMMARY

- Most managers rely on a limited subset of the Human Resources Effectiveness processes available.

- Performance is a combination of what people do (their behavior) and the results they achieve (their accomplishments or outputs).

- You can try to influence behavior *before* it happens or *after* it occurs.

- The things you use to try to influence what people do before they do it are called antecedents.

- When you try to influence behavior *after* it takes place, you provide "consequences."

- Remember your ABCs—an **A**ntecedent prompts a **B**ehavior, which is followed by a **C**onsequence.

- Managers spend too much time on antecedents and not enough time providing consequences, yet consequences are the single most effective tool a manager has for increasing employee performance and improving morale (Daniels, p. 23).

- The three main characteristics of consequences that determine the influence they have on behavior are:
  — Whether the *performer* views the consequence as positive or negative.
  — The timing (immediate or future) of the consequence.
  — The certainty of the consequence.

- ABC analysis is the process of systematically examining all the antecedents and consequences associated with a given behavior or performance in order to better understand why people do what they do (Daniels, p. 37).

- The balance of consequences worksheet (see Chapter 11) is a tool you can use to help you perform an ABC analysis.

- Refer to *Performance Management,* by Aubrey Daniels, for more information on the ABC model of behavior change, or ABC analysis.

HUMAN RESOURCES EFFECTIVENESS ACTIONS CHECKLIST

Stage 1

1. Identify the current balance of consequences associated with improvement actions for each of your associates.

2. Identify whether the organization's existing Human Resources Effectiveness processes are "user friendly."

3. Reward progress as well as results.

4. Celebrate small and large victories.

5. Create development plans for your associates that include the skills needed to improve customer satisfaction.

6. Identify the factors that cause and hinder employee satisfaction.

7. Measure your employees' satisfaction, especially their perception of you as internal supplier.

8. Establish improvement goals to improve employee satisfaction.

9. Provide training for your associates in the use of quality processes and tools.

10. Tailor rewards/recognition to the individual.

11. Identify barriers to effective communication.

12. Assess the effectiveness of feedback associates receive.

13. Revise new-hire orientation and training to ensure that training on QMS values and the improvement processes are included.

14. Schedule training for an associate.

15. Initiate a project to increase the effectiveness of feedback associates receive.

16. Provide time for improvement teams to meet.

17. Recognize employees who participate in improvement projects.

18. Protect (as in "don't shoot") the messenger!

19. Reward open, honest communication.

20. Provide incentives for associates to use quality tools.

21. Remove barriers that prevent associates from using quality tools.

Stage 2

1. Align the balance of consequences for everyone in your area so that quality improvement is encouraged.

2. Learn from mistakes and failures—don't punish team members for trying and failing. Show your appreciation of their efforts and encourage them to try again.

3. Initiate projects to increase the user friendliness of the organization's Human Resources Effectiveness processes.

4. Design jobs so they provide built-in feedback.

5. Design jobs so they provide built-in positive, immediate, and certain consequences.

6. Examine the collective (combined) effect of Human Resources Effectiveness processes on quality improvement—do they complement one another and reinforce quality, or do they contradict one another and send a mixed or incorrect quality improvement message?

Stage 3

1. Link compensation to customer satisfaction and retention data.

2. Promote people based on their quality improvement track record.

3. Shift more responsibility for the processes and outcomes to employees and away from the managers. This could include improvement activities all the way up to self-directed work teams.

4. Establish reward/recognition plans that reinforce quality improvement (e.g., link pay directly to customer satisfaction).

5. Align the collective (combined) effect of Human Resources Effectiveness processes on quality improvement so they complement one another and reinforce quality.

SECTION 3

CHAPTER ELEVEN

THE MANAGER'S TOOLKIT

EXECUTIVE OVERVIEW

- This chapter contains three tools designed to help you put quality into practice:

  — Personal Quality Planner
  — Personal Quality Checksheet
  — Balance of Consequences Worksheet

- The Personal Quality Planner (also known as the Quality Management System Planning Worksheet) is a weekly time management planning tool that helps you plan and schedule the actions you wish to take to implement each of the seven quality management system (QMS) elements.

- The Personal Quality Checksheet is a specially designed checksheet you can use to track any QMS-related action, event, or process that you feel is especially important to monitor.

- The Personal Quality Planner and the Personal Quality Checksheet are often used together; both are designed to help you make use of the items listed in the Manager Action Checklists that appear at the end of Chapters 4–10.

- The Balance of Consequences Worksheet helps you analyze and align the balance of consequences for your associates as part of your Human Resources Effectiveness processes. Please refer to Chapter 10, "Human Resources Effectiveness," to learn more about this important tool.

Quality Management System Planning Worksheet

Quality Management System Element/Planned Actions

	Time Period →	Monday	Tuesday	Wednesday	Thursday	Friday	Weekend
Leadership							
Management By Fact							
Human Resources Effectiveness							
Total Customer Satisfaction							
Quality Results							
Strategic Quality Planning							
Process Quality Assurance							

These Elements/Actions transcend *every* work process

These Elements/Actions should be directed at one work process at a time

1
2
3
4

Action(s) taken directly impact customer satisfaction — 20

Action(s) taken to align the balance of consequences to reinforce quality improvement — 30

Provides positive, immediate, certain consequence — 20
Visible by others — 10

Action(s) taken to communicate the quality message — 10

Words match deeds — 5
Message is consistent — 5
Content relevant to (tailored to match job of) each associate — 5

Action(s) taken to visibly spend time improving quality — 20

Includes a Management By Fact action — 10
Includes action that aligns consequences to reinforce quality improvement — 10
Includes use of statistical tools to reach conclusions — 10
Actions demonstrate knowledge of type of variation present — 10
Action is part of a root-cause analysis — 10

Action(s) taken for all three key principles — 20

Action(s) taken for all seven management system elements — 20

Action(s) taken for "big three" management system elements — 20

Three Key Principles:
• Integrate Quality Message into Communication Daily
• Visibly Spend Time Daily Improving Quality
• Align the Balance of Consequences to Reinforce Quality Improvement

1. Determine which of the three stages of implementation best characterizes your organization, department, or group right now.
2. Define the goal you plan to achieve for each management system element this week.
3. Refer to the Manager Action Checklists for each management system element to see a list of actions that match each of the three stages of implementation.
4. Schedule the actions you will take to achieve each of your goals.
5. List any actions that you wish to monitor more closely on your Personal Quality Checksheet.
6. Strive to reach a daily point total of 100.

FIGURE 11.1

WHAT IS A PERSONAL QUALITY PLANNER?

The Personal Quality Planner is a worksheet designed to help you spend your time on each of the seven elements that make up a QMS. See Figure 11.1.

WHEN SHOULD I USE
THE PERSONAL QUALITY PLANNER?

The Personal Quality Planner should be used to plan your quality management actions weekly. If you use some sort of time management system, try to incorporate this tool into your planning and scheduling activities. The actions listed on your worksheet should be treated as "A" priority items or "Q2" (high-importance) activities.

HOW TO COMPLETE
A PERSONAL QUALITY PLANNER

The example in Figure 11.2 shows a partially completed Personal Quality Planner. The numbered items illustrate some of the six steps below. The worksheet contains a Stage 1 leadership action that the manager has chosen to help his staff use a quality tool. Refer to Table 11.1 and score this manager's planned action. (*Hint:* It's probably more than you think.)

WANT PRACTICE USING THIS TOOL?

Complete a Personal Quality Planner based on the following information. Check your understanding by referring to the completed worksheet in Figure 11.3.

John is planning to start an improvement project in the area he manages. His company has collected and analyzed customer data and used it to identify 10 key processes that impact customer satisfaction, though process owners have not yet been identified. The firm established overall cycle time and defect reduction improvement objectives for those processes and has briefed all the managers involved (John is one of those managers). John has just attended training in continuous improvement as well as a course titled, "How to Effectively Sponsor an Improvement Project." As John reviews the data the firm collected and

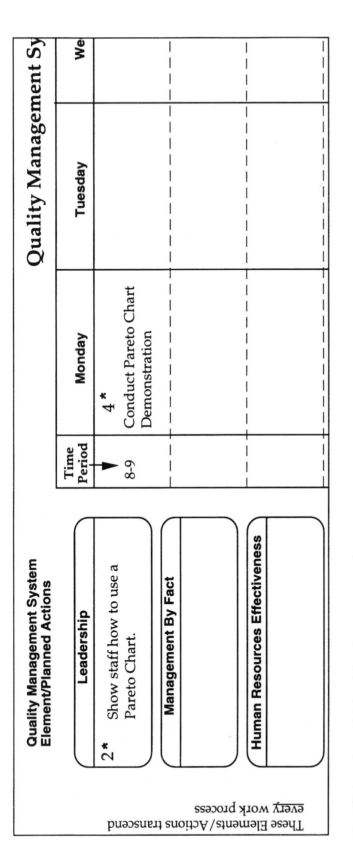

Quality Management Sy

1. Determine which of the three stages of implementation best characterizes your organization, department, or group, right no (i.e., Awareness Assessment, Integration, or Maintenance).

2* Define the goal you plan to achieve for each management system element this week.

3. Refer to the Manager Action Checklists for each management system element to see a list of actions that match each of the three stages of implementation.

4* Schedule the actions you will take to achieve each of your goals.

5. List any actions you wish to monitor more closely on your Personal Quality Checksheet.

6. Strive to reach a daily point total of 100. (see Table 11.1).

FIGURE 11.2

TABLE 11.1

Daily Point Scoring Table

Action(s) taken directly impact customer satisfaction	20
Action(s) taken to align the balance of consequences to reinforce quality improvement	30
— Provides positive, immediate, certain consequence	20
— Visible by others	10
Action(s) taken to communicate the quality message	10
— Words match deeds	5
— Message is consistent	5
— Content relevant to (tailored to match job of) each associate	5
Action(s) taken to visibly spend time improving quality	20
— Includes a Management by Fact action	10
— Includes action that aligns consequences to reinforce quality improvement	10
— Includes use of statistical tools to reach conclusions	10
— Actions demonstrate knowledge of type of variation present	10
— Action is part of a root cause analysis	10
Action(s) taken for all three key principles	20
Action(s) taken for all seven management system elements	20
Action(s) taken for "big three" management system elements	20

the resultant improvement goals, his thoughts turn to improvement team member selection. He knows that it is not enough to simply appoint a team, send them to training, and let the project take care of itself. He wants to get the project off to a good start, so he schedules the training so that it will be completed just prior to starting the improvement project. Then, he plans to sit down with each team member individually prior to the training, explain the upcoming improvement project, and point out how the training will help prepare the associate to successfully participate in the project. He decides to anticipate and address any concerns team members may have by identifying how they will view working on the project based on his understanding of what makes each tick. He also plans a series of actions he will take for each team member to encourage and reward the effort each demonstrates upon returning from the training. *Hint:* **Assume that John schedules the training now though it will be conducted later, and that his planning for each team member is part of a single broader action.**

Quality Management System Planning Worksheet

Quality Management System Element/Planned Actions

Element	Time Period	Monday	Tuesday	Wednesday	Thursday	Friday	Weekend
Leadership							
Management By Fact							
Human Resources Effectiveness	11	Schedule training					
Prepare project team							
Total Customer Satisfaction							
Quality Results	3-4		Analyze Balance of Consequences				
Strategic Quality Planning							
Process Quality Assurance							

These Elements/Actions transcend every work process

These Elements/Actions should be directed at one work process at a time

Three Key Principles:
- Integrate Quality Message into Communication Daily
- Visibly Spend Time Daily Improving Quality
- Align the Balance of Consequences to Reinforce Quality Improvement

1. Determine which of the three stages of implementation best characterizes your organization, department, or group right now.
2. Define the goal you plan to achieve for each management system element this week.
3. Refer to the Manager Action Checklists for each management system element to see a list of actions that match each of the three stages of implementation.
4. Schedule the actions you will take to achieve each of your goals.
5. List any actions that you wish to monitor more closely on your Personal Quality Checksheet.
6. Strive to reach a daily point total of 100.

Action(s) taken directly impact customer satisfaction	20
Action(s) taken to align the balance of consequences to reinforce quality improvement	30
Provides positive, immediate, certain consequence	20
Visible by others	10
Action(s) taken to communicate the quality message	10
Words match deeds	5
Message is consistent	5
Content relevant to (tailored to match job of) each associate	5
Action(s) taken to visibly spend time improving quality	20
Includes a Management By Fact action	10
Includes action that aligns consequences to reinforce quality improvement	10
Includes use of statistical tools to reach conclusions	10
Actions demonstrate knowledge of type of variation present	10
Action is part of a root-cause analysis	10
Action(s) taken for all three key principles	20
Action(s) taken for all seven management system elements	20
Action(s) taken for "big three" management system elements	20

FIGURE 11.3

John's organization is at Stage 1. He has chosen two Stage 1 actions to help achieve his Human Resources Effectiveness goal of preparing the improvement project team. He has scheduled these actions for Monday and Tuesday. We'll assume that he has decided he does not need to monitor these actions more closely using a Personal Quality Checksheet. John's point total may be calculated as follows: Actions taken to align the balance of consequences to reinforce quality improvement = 30.

The point total would be more if you concluded that this actions would directly impact customer satisfaction (based on the company's earlier collected data).

Don't worry if your point totals don't match the ones shown here; the important thing is that you better understand how this tool can help you put quality into practice.

WHAT IS A PERSONAL QUALITY CHECKSHEET?

A Personal Quality Checksheet is a tool designed to help you track opportunities and missed opportunities associated with a process, event, or action.

The Personal Quality Checksheet presented in Figure 11.4 an adaption of the personal quality checklist described by Dr. Harry Roberts of the University of Chicago. (For more detailed information, please refer to *Quality Is Personal,* by Harry V. Roberts and Bernard F. Sergesketter.)

OPPORTUNITY VS. MISSED OPPORTUNITY

As you can see from the example in Figure 11.4, opportunities are instances of taking a given action, such as to reviewing improvement project progress, recognizing team progress, obtaining customer expectations data, etc. A missed opportunity is any of these same instances that do *not* occur (i.e., failing to review improvement project progress or not recognizing team progress when you have the team together, etc.).

WHEN SHOULD I USE
A PERSONAL QUALITY CHECKSHEET?

Use the Personal Quality Checksheet whenever you want to concentrate or focus your attention on an important process or action. You can also use the checksheet to reduce the number of errors or missed opportunities within

PERSONAL QUALITY CHECKSHEET

QUALITY MANAGEMENT SYSTEM PROCESS	DAY OF THE MONTH →	1	2	3	4	
Review improvement project progress	opportunities	1	0	2	0	
	missed opportunities	0	0	1	0	
Recognize team progress	opportunities	0	1	0	0	
	missed opportunities	0	0	0	0	
Obtain customer expectations data for three processes	opportunities	1	1	1	0	
	missed opportunities	0	0	0	0	
Communicate the importance of quality improvement in meetings	opportunities	3	1	2	4	
	missed opportunities	0	1	1	0	

FIGURE 11.4

your control, to measure your current level of performance, and to track your improvement.

Unlike the Personal Quality Planner, which is a weekly tool, the Personal Quality Checksheet can be used according to the process or action chosen (i.e., you may not have daily opportunities to review project progress or recognize team progress, but you definitely want to take advantage of these opportunities whenever they do occur).

HOW TO COMPLETE
A PERSONAL QUALITY CHECKSHEET:

Figure 11.5 shows a partially completed Personal Quality Checksheet. The numbered items illustrate some of the four steps below the figure. The worksheet shows two actions that the manager wishes to closely monitor because of their importance. During the first four days of the month, there are a total of three complete opportunities to review improvement project progress, and one to recognize the progress of the team. Note that the manager missed one opportunity to review improvement project progress.

PERSONAL QUALITY CHECKSHEET

QUALITY MANAGEMENT SYSTEM PROCESS ↓ DAY OF THE MONTH →		1	2	3	4		4.* Totals
Review improvement project progress 1.*	opportunities 2.*	1	0	2	0		3
	missed opportunities 3.*	0	0	1	0		1
Recognize team progress	opportunities	0	1	0	0		1
	missed opportunities	0	0	0	0		0

1.* Identify the process, event, or action you would like to monitor and improve.
2.* Record the number of opportunities for each process, event, or action.
3.* Record the number of missed opportunities.
4.* Collect data for at least two months to establish your baseline performance.

FIGURE 11.5

WANT PRACTICE USING THIS TOOL?

Complete a Personal Quality Checksheet based on the following information. Check your understanding by referring to the completed checksheet in Figure 11.6.

Phyllis has completed a Personal Quality Planner and wants to closely monitor the following actions: integrate a quality message into all her formal communications, demonstrate that starting meetings on time is important, and increase her use of quality tools, especially when it's visible to others. She has defined formal communications as, "written and planned verbal communication." Here is some additional information on her upcoming week, May 5–9:

- The weekly report is due Thursday.

- Two presentations are planned: One on Tuesday, the other on Friday.

- The staff meeting is Wednesday, as usual.

- She has a meeting scheduled for Tuesday to work with one of the improvement project teams; the team is trying to analyze defect data collected using a checksheet. She thinks a Pareto Chart may help.

At the end of the week, Phyllis noted that she had included the results of the improvement project team meeting in the weekly report, added a quality

PERSONAL QUALITY CHECKSHEE

QUALITY MANAGEMENT SYSTEM PROCESS	DAY OF THE MONTH →	1	2	3	4	5	6	7	8	9	10	11	12	13	14	15	16	17	18
Integrate quality message into formal communication.	opportunities					0	1	0	1	1									
	missed opportunities					0	1	0	0	0									
Start meetings on time.	opportunities					0	1	1	0	0									
	missed opportunities					0	0	0	0	0									
Use quality tools, especially when visible.	opportunities					0	1	0	0	0									
	missed opportunities					0	0	0	0	0									
	opportunities																		
	missed opportunities																		
	opportunities																		
	missed opportunities																		

FIGURE 11.6   Phyllis's Personal Quality Checksheet

topic to the agenda for the staff meeting, and discovered that the improvement project team did indeed find the Pareto Chart to be of help in analyzing the data it had collected. She looked for ways to build a quality message into her two presentations but was unable to do it for the Tuesday presentation. She started all her meetings on time.

WHAT IS A BALANCE OF CONSEQUENCES WORKSHEET?

The Balance of Consequences worksheet (see Figure 11.7) is a tool designed to help you do an ABC analysis. (Refer to Chapter 10 for information on how to do an ABC analysis. You may also wish to read *Performance Management,* by Aubrey Daniels, who developed the ABC analysis technique.) ABC analysis will help you learn how you can align the balance of consequences that each of your associates experiences so that quality improvement is strongly reinforced.

WHEN SHOULD I USE A BALANCE OF CONSEQUENCES WORKSHEET?

Use a Balance of Consequences Worksheet for each associate whenever you want to discover ways to make his or her job more satisfying or rewarding. It is especially helpful if you are looking for ways to match work process requirements to the personal values of associates.

HOW TO COMPLETE A BALANCE OF CONSEQUENCES WORKSHEET

The worksheet in Figure 11.8 has been completed for a person who drives faster than the posted speed limit on his way to and from work. Refer to Chapter 10, pages 124–131, to see how this example was created.

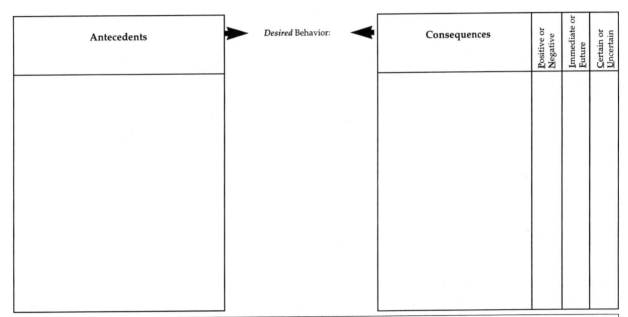

Name:

Antecedents	→	*Desired* Behavior:	←	Consequences	Positive or Negative	Immediate or Future	Certain or Uncertain

Instructions:
1. Complete two worksheets for <u>each</u> individual (start with the *undesired* behavior and then complete a worksheet for the **desired** behavior).
2. Write the person's name and the **desired** behavior for whom you are doing this analysis.
3. List all the possible antecedents that come to mind associated with the **desired** behavior.
4. List all the possible consequences that come to mind associated with the **desired** behavior.
5. Review the list of consequences and mark out those that are relevant to the company rather than the performer.
6. Classify the remaining consequences as P/N, I/F, or C/U.
7. Review the antecedents and consequences currently in place. Does the balance (mix) reinforce what you want it to? If not, align the balance of consequences; add Ps, P/I's & Is and antecedents for the **desired** behavior.

FIGURE 11.7   Balance of Consequences Worksheet

WANT PRACTICE USING THIS TOOL?

Complete the Balance of Consequences worksheets for *Chris* based on the information below. Check your understanding by referring to the completed worksheets in Figures 11.9 and 11.10. This information is based on the explanation provided in Chapter 10. You will find it much easier to use this tool if you read that chapter first.

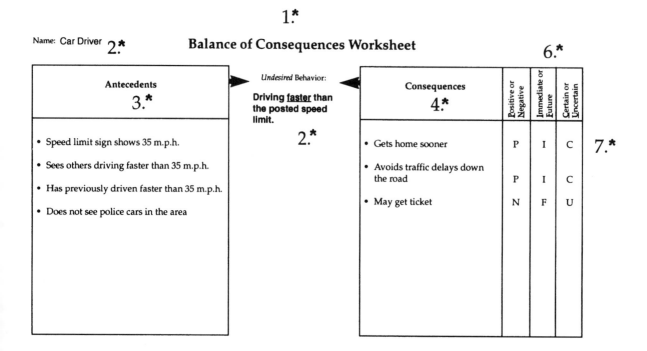

1.*

Name: Car Driver   2.*   **Balance of Consequences Worksheet**   6.*

Antecedents 3.*	Undesired Behavior: Driving **faster** than the posted speed limit. 2.*	Consequences 4.*	Positive or Negative	Immediate or Future	Certain or Uncertain	
• Speed limit sign shows 35 m.p.h. • Sees others driving faster than 35 m.p.h. • Has previously driven faster than 35 m.p.h. • Does not see police cars in the area		• Gets home sooner	P	I	C	7.*
		• Avoids traffic delays down the road	P	I	C	
		• May get ticket	N	F	U	

1.* Complete two worksheets for <u>each</u> individual (start with the *undesired* behavior and then complete a worksheet for the **desired** behavior).

2.* Write the person's name and the *undesired* behavior for whom you are doing this analysis.

3.* List all the possible antecedents that come to mind associated with the *undesired* behavior.

4.* List all the possible consequences that come to mind associated with the *undesired* behavior.

5.* Review the list of consequences and mark out those that are relevant to the company rather than the performer.

6.* Classify the remaining consequences as PN, IF, or CU (see chapter 10 for explanation).

7.* Review the antecedents and consequences currently in place-does the balance (mix) reinforce what you want it to? If not, align the balance of consequences-add Ps & Is and antecedents for the **desired** behavior.

FIGURE 11.8

Consider the following situation:

You are the manager of an assembly area in a plant that makes telecommunication products. You have 15 people that report to you, all of whom are skilled technicians with varying years of service. Your company has established a continuous improvement process and all your people have attended training in how the improvement process works. You've also attended the training, and buy in to the fact that unless the company makes continuous improvement a way of life, competitive forces in the industry will continue to cause downsizing, consolidation, and all the other things that put your job at risk. Besides, you personally believe that increasing customer satisfaction is the right thing to do, and you want to do your part. Accordingly, you study customer satisfaction data and decide to start an improvement project to reduce the number of defects associated with the assemblies your area produces. You create a team made up of three people who work on the assembly process: George Johnson, Mary Adams, and Chris Cobb. You call the three of them together, review the customer satisfaction data with them, and challenge them to use the company's improvement process to find and eliminate the root causes of the defects. You explain to the team that this is an important project, not just because it is the first one to be done in your area, but because the assemblies made here are vital to the overall profitability of one of the firm's major product lines. You establish a regular weekly schedule to meet with the team to discuss progress or otherwise work with them as they see fit. Near the end of the meeting, you ask, "Are there any questions about this project?" The three of them look at each other, then at you. Finally, George says, "Frank over in building C has been working on a quality project. He says that it puts a lot of pressure on him, because he's still responsible for his regular work. Is that the case here too?" Then, Chris speaks up and says, "As you know, my wife just had a baby, our first. We both work and we've agreed to share the child-rearing responsibilities equally. If this project means overtime, I can't do it, even if it's at time and a half." Mary then says, "I like the idea of trying to make things better, I really do. But it's been over six months since I've attended that quality training course you sent me to. I'm afraid I'm a little rusty on what they covered." "Me too," the others chime in. "Is there anyone we can turn to for help if we have any questions?"

"Hold on," you say. "Let's take these one at a time. Now, about the work load . . ."

Balance of Consequences Worksheet

Name: Chris Cobb

Antecedents		Consequences	Positive or Negative	Immediate or Future	Certain or Uncertain
• Hears peers talk about how much time it takes		• Avoids getting behind on regular work	P	I	C
• Concerned about not knowing how to follow the continuous improvement process		• Avoids looking ignorant in front of peers	P	I	C
• Sees others work late to catch up with normal work load		• Avoids overtime	P	I	C
• No one else in his area is doing it		• Might negatively affect next performance appraisal	N	F	U
• New addition to the family		• Might get a lecture on being a team player	N	F	U

Undesired Behavior:

Not working on improvement project.

Instructions:

1. Complete two worksheets for <u>each</u> individual (start with the *undesired* behavior and then complete a worksheet for the **desired** behavior).
2. Write the person's name and the **desired** behavior for whom you are doing this analysis.
3. List all the possible antecedents that come to mind associated with the **desired** behavior.
4. List all the possible consequences that come to mind associated with the **desired** behavior.
5. Review the list of consequences and mark out those that are relevant to the company rather than the performer.
6. Classify the remaining consequences as P/N, I/F, or C/U.
7. Review the antecedents and consequences currently in place. Does the balance (mix) reinforce what you want it to? If not, align the balance of consequences; add Ps & Is and antecedents for the **desired** behavior.

FIGURE 11.9

158

Balance of Consequences Worksheet

Name: Chris Cobb

Desired Behavior:

Working on improvement project.

Antecedents	Consequences	Positive or Negative	Immediate or Future	Certain or Uncertain
• Regular weekly meetings with the boss	• Attending weekly meetings will take time away from other things	N	I	C
• Training on the continuous improvement process	• May have to attend more training, which will take more time	N	I	C
• High-visibility project	• Working on this project may be risky	N	I	C
	• People that don't know me will be judging my work on this project	N	F	C
	• Project may make things better	P	F	U

Instructions:

1. Complete two worksheets for each individual (start with the *undesired* behavior and then complete a worksheet for the **desired** behavior).

2. Write the person's name and the **desired** behavior for whom you are doing this analysis.

3. List all the possible antecedents that come to mind associated with the **desired** behavior.

4. List all the possible consequences that come to mind associated with the **desired** behavior.

5. Review the list of consequences and mark out those that are relevant to the company rather than the performer.

6. Classify the remaining consequences as P/N, I/F, or C/U.

7. Review the antecedents and consequences currently in place. Does the balance (mix) reinforce what you want it to? If not, align the balance of consequences; add Ps & Is and antecedents for the **desired** behavior.

FIGURE 11.10

159

SUMMARY

- Your manager's toolkit consists of three tools designed to help you put quality into practice:

  — Personal Quality Planner
  — Personal Quality Checksheet
  — Balance of Consequences Worksheet

- Use the Personal Quality Planner when you want to plan and schedule the actions you wish to take to implement each of the seven QMS elements.

- Whenever you want to monitor a particularly important action more closely, use the Personal Quality Checksheet.

- Both the Personal Quality Planner and the Personal Quality Checksheet are designed to help you make use of the items listed in the Manager Action Checklists that appear at the end of Chapters 4–10.

- If you want to strengthen your Human Resources Effectiveness processes, learn how to use the Balance of Consequences Worksheet. (See Chapter 10, "Human Resources Effectiveness," to learn more about this powerful tool.

- Blank copies of each of these tools follow this page. They may be photocopied for personal use only.

Quality Management System Planning Worksheet

Quality Management System Element/Planned Actions

	Time Period →	Monday	Tuesday	Wednesday	Thursday	Friday	Weekend
Leadership							
Management By Fact							
Human Resources Effectiveness							
1 **Total Customer Satisfaction**							
2 **Quality Results**							
3 **Strategic Quality Planning**							
4 **Process Quality Assurance**							

These Elements/Actions transcend every work process

These Elements/Actions should be directed at one work process at a time

Three Key Principles:
- Integrate Quality Message into Communication Daily
- Visibly Spend Time Daily Improving Quality
- Align the Balance of Consequences to Reinforce Quality Improvement

1. Determine which of the three stages of implementation best characterizes your organization, department, or group right now.

2. Define the goal you plan to achieve for each management system element this week.

3. Refer to the Manager Action Checklists for each management system element to see a list of actions that match each of the three stages of implementation.

4. Schedule the actions you will take to achieve each of your goals.

5. List any actions that you wish to monitor more closely on your Personal Quality Checksheet.

6. Strive to reach a daily point total of 100.

Action(s) taken directly impact customer satisfaction	20
Action(s) taken to align the balance of consequences to reinforce quality improvement	30
Provides positive, immediate, certain consequence	20
Visible by others	10
Action(s) taken to communicate the quality message	10
Words match deeds	5
Message is consistent	5
Content relevant to (tailored to match job of) each associate	5

Action(s) taken to visibly spend time improving quality	20
Includes a Management By Fact action	10
Includes action that aligns consequences to reinforce quality improvement	10
Includes use of statistical tools to reach conclusions	10
Actions demonstrate knowledge of type of variation present	10
Action is part of a root-cause analysis	10
Action(s) taken for all three key principles	20
Action(s) taken for all seven management system elements	20
Action(s) taken for "big three" management system elements	20

PERSONAL QUALITY CHECKSHEET

QUALITY MANAGEMENT SYSTEM PROCESS	DAY OF THE MONTH →	1	2	3	4	5	6	7	8	9	10	11	12	13	14	15	16	17	18	19	20	21	22	23	24	25	26	27	28	29	30	31
	opportunities																															
	missed opportunities																															
	opportunities																															
	missed opportunities																															
	opportunities																															
	missed opportunities																															
	opportunities																															
	missed opportunities																															
	opportunities																															
	missed opportunities																															

1. Identify the process, events, or actions you would like to monitor and improve.

2. Record the number of opportunities for the process, event, or action.

3. Record the number of missed opportunities.

4. Collect data for at least two months and use it to establish your current performance level (baseline).

Balance of Consequences Worksheet

Name:

Antecedents		*Undesired* Behavior:	Consequences	Positive or Negative	Immediate or Future	Certain or Uncertain

Instructions:

1. Complete two worksheets for each individual (start with the *undesired* behavior and then complete a worksheet for the **desired** behavior).
2. Write the person's name and the **desired** behavior for whom you are doing this analysis.
3. List all the possible antecedents that come to mind associated with the **desired** behavior.
4. List all the possible consequences that come to mind associated with the **desired** behavior.
5. Review the list of consequences and mark out those that are relevant to the company rather than the performer.
6. Classify the remaining consequences as PN, IF, or CU.
7. Review the antecedents and consequences currently in place. Does the balance (mix) reinforce what you want it to? If not, align the balance of consequences; add Ps & Is and antecedents for the **desired** behavior.

Name:

Balance of Consequences Worksheet

Antecedents	Desired Behavior:	Consequences	Positive or Negative	Immediate or Future	Certain or Uncertain

Consequences

Instructions:

1. Complete two worksheets for each individual (start with the *undesired* behavior and then complete a worksheet for the **desired** behavior).
2. Write the person's name and the **desired** behavior for whom you are doing this analysis.
3. List all the possible antecedents that come to mind associated with the **desired** behavior.
4. List all the possible consequences that come to mind associated with the **desired** behavior.
5. Review the list of consequences and mark out those that are relevant to the company rather than the performer.
6. Classify the remaining consequences as PN, IF, or CU.
7. Review the antecedents and consequences currently in place. Does the balance (mix) reinforce what you want it to? If not, align the balance of consequences; add Ps & Is and antecedents for the **desired** behavior.

INDEX